Tattoos in American Visual Culture

Tattoos in American Visual Culture

Mindy Fenske

palgrave
macmillan

TATTOOS IN AMERICAN VISUAL CULTURE

First published in 2007 by
PALGRAVE MACMILLAN™
175 Fifth Avenue, New York, N.Y. 10010 and
Houndmills, Basingstoke, Hampshire, England RG21 6XS
Companies and representatives throughout the world.

PALGRAVE MACMILLAN is the global academic imprint of the Palgrave Macmillan division of St. Martin's Press, LLC and of Palgrave Macmillan Ltd. Macmillan® is a registered trademark in the United States, United Kingdom and other countries. Palgrave is a registered trademark in the European Union and other countries.

ISBN-13: 978–0–230–60027–0

Library of Congress Cataloging-in-Publication Data

Fenske, Mindy, 1969–
Tattoos in American visual culture / Mindy Fenske.
p. cm.
Includes bibliographical references.
ISBN 0–230–60027–1
1. Tattooing 2. Tattooing—Social aspects. 3. Body image. I. Title.

GT2345.F455 2007
391.6′5—dc22 2007010215

A catalogue record for this book is available from the British Library.

Design by Newgen Imaging Systems (P) Ltd., Chennai, India.

First edition: December 2007

10 9 8 7 6 5 4 3 2 1

Printed in the United States of America.

Transfererd to Digital Printing in 2009

For Mom

Contents

List of Illustrations

Acknowledgments

Throughout the process of conceiving, researching, writing, and obsessing about this project, I have been a grateful and undeserving recipient of encouragement, intellectual support, and basic nourishment from many groups and individuals. I want to first extend my gratitude to colleagues in the Department of English at the University of South Carolina who generously either read and critiqued this work, or, equally valuably, offered me their critical ear at various stages of its development. I especially want to acknowledge the guidance and support of Rebecca Stern, John Muckelbauer, Susan Courtney, Dan Smith, Kristan Poirot, Pat Gehrke, Holly Crocker, and Tony Jarrells.

At the beginning of this project and my professional career, I was happily ensconced in heart of South Louisiana. Members of my Baton Rouge family have dispersed but continue to sustain me in heart and mind. For their never flinching, ever truthful, all too and none too serious friendship and advice, special thanks to Gretchen Stein Rhodes, Danielle Sears Vignes, Lisa Swartzel Flanagan, Gary Reeves, Shaun Treat, Dan Grano, and Justin Trudeau.

Members and friends of the performance studies community across the country have also supported my work both in material and nonmaterial ways. Michael Bowman, Ruth Laurion Bowman, and Trish Suchy encouraged this project from the time it was a mere idea. Joshua Gunn has been instrumental in providing a role model for professionalism both as a scholar and in how to serve the profession and honor its individuals with grace and generosity. Marcyrose Chvasta, Michael Levan, and the entire Underscore Performance Collective have challenged my thought, encouraged me to be creative, and fostered in me a sense of social advocacy (and we have had a beer or two along the way).

I would like to extend my gratitude to the College of Liberal Arts (now College of Arts and Sciences) and the Department of English at the University of South Carolina. A CLASS grant from the College and Summer Stipend from the department supported the ethnographic fieldwork. Thanks also to the National Communication Association and Taylor and Francis Publishing. A previous version of chapter two appeared as "Movement and

Resistance: (Tattooed) Bodies and Performance," *Communication and Critical/Cultural Studies* 4.1 (March 2007): 51–73.

Thanks to The Special Collections Library at Syracuse for permission to reprint photographs from the Ronald G. Becker Collection of Charles Eisenmann Photographs and to the National Archives for the Rosie the Riveter poster.

Finally, without the humor and support of my family, none of this would be possible. Pop and Scott, you know me better than most and yet you still invite me home for the holidays! To my partner in life, Jason Munsell, my love and gratitude for your patience and unflagging support.

INTRODUCTION: PICTURING TATTOOS

Representations are bodies too![1]

SINCE THE RISE OF THE DIME MUSEUM, tattooed bodies have been displayed and exhibited in American culture. In the late nineteenth and early twentieth centuries the carnival sideshow and "ten-in-one" freak stages of P.T. Barnum and others featured "illustrated" men and women. In the early to mid-twentieth century, sideshow popularity faded with the advent of radio and television. At about the same time, audience fascination with human "freaks" turned into wary disgust. Partly in response to the new entertainment technologies and the alteration in attitudes toward freaks, the illustrated person moved from the carnival freak stage to mass-mediated venues such as advertising and film. Tattoos also shifted from cultural displays of freakery into displays of masculinity, military group affiliation, and the working class. Since the late 1960s, tattooing has apparently moved again. One of the hallmarks of the post-1960s tattoo renaissance has been the increasing popularity of tattooing for women, celebrities, and members of the middle class. Tattooed bodies in the early twenty-first century frequently appear in advertisements, on television programs, and in films. Tattooed bodies also parade across stages at tattoo conventions, at sports arenas, in performance art venues, and at rock concerts. Even Barbie has had a tattoo.

What is most fascinating about these practices of displaying tattoos concerns the power of visual images of the tattooed body. In particular, what is intriguing to me is the power of these (and other) images to destabilize, disrupt, and potentially resist ideological discourses that seek to control and contain what they mean. This book is therefore about how meaning has been made by/at the site of the image of the tattooed body in U.S. culture throughout the twentieth century. Throughout, the focus is on how meanings are discursively constructed and historically deployed through, upon, and by the tattooed human body in contexts of visual display. In particular, this book is interested in the specific discursive structures through which gender, class, and exoticism are produced at the site of the tattooed body on display.

As anyone who has ever gotten a tattoo can attest, the display of the tattoo can incite a variety of responses depending on a combination of factors including (but not limited to) the social context of display, the gender-identification of the tattooed person, the location and size of the tattoo, and the style and artistry of the tattoo. For example, on more than one occasion, people who have seen my tattoos have made the comment, "I would have never taken *you* for someone who has tattoos." Apparently my white, female, middle-class college professor status (along with my runner's tank top and shorts, strawberry-blond ponytail, and Nike shoes) does not always neatly fall into visual place with some of the various images of tattooed people that have been and are being produced in popular U.S. culture. If and when I tell people that I come from a working-class background, that my dad used to drive a motorcycle, and that I study performance and popular culture, however, a look of slow comprehension generally replaces the confused countenance. The attribution of meaning made possible by the history of class, gender, and cultural images associated with categories such as "working class" and "performance art" allow certain people to at least temporarily make sense out of the image of my tattooed body. Throughout the book, I attend, therefore, to the specific structures of the discourses of class, gender, and exoticism circulating around and through visual representations of tattooed bodies.

I am also not convinced that the image of (my) tattooed body is held completely still by these categories of evaluation. While it is certainly true that visual images of tattooed bodies are often socially inscribed by class and gender evaluations, it is also the case that the inscriptions do not have the power to fully contain the visual image. My tattooed body absent the verbal personal history, for instance, still has the power to confuse. This book is therefore also about the power of the visual image to act out of and through the discursive structures that seek to contain its meaning. The power to, in other words, perform in and through normative discourse. The purpose is to describe the environment of discursive structures through which gender, class, and exotic meaning have been produced in order to show the flashes that occur when visual images act (perform) in some way to destabilize, disrupt, and potentially resist the forces of normative evaluation. There are moments when, for instance, images of tattooed bodies actually ask viewers to question taken-for-granted assumptions about class or gender even as the display of those bodies simultaneously reinforces those assumptions. It is these moments of destabilization and disruption that enact the agency, the power to act and effect change, within and of images. The central thesis of the book is that flashes of instability in meaning (agency) occur because of and through the performance of discursive structures and not only in spite of or in resistance to them. The active force in visual representation, moreover, depends upon the specific mobilization and character of the discursive environment through which meanings are made. In other words, the trouble that people sometimes have in making sense out of my tattooed image is both productive and absolutely dependent upon the fact that I come across as the average strawberry blonde, white, middle-class, thirty-something, Midwestern U.S., athletic female.[2]

At this point, it is probably becoming clear that this book is as much about the agency and performance of images as it is about tattooed bodies. That being said, the choice of tattoos is neither accidental nor arbitrary. The choice to concentrate on the tattooed body in this study of the visual image is initially important because of the roles history and context play in understanding agency. I argue that the agency of the image is built through the specific social, cultural, and historical discourses that it operates through and within. That is, in order to ascertain how gender or class operates within any given visual image, and how the image acts out of gender and class, it is necessary to attend to the variety of cultural and historical forces circulating through that specific image. This makes the choice of a particular sight/site of analysis crucial for the analysis of agency. It is only through actual image locations, and their particular histories and discourses, that agency happens.

The selection of this precise sight/site is additionally warranted because of the unique visual power of the tattooed body. Tattooed bodies continue to act as a catalyst for a variety of social anxieties about the body and display. Tattooed bodies call hyperbolic attention to the ways in which meanings are made in and through images of the human body. Tattooed bodies bring into sharp metaphorical relief ideas about, for instance, the inscription of cultural and social power upon bodies (tattooed bodies are, after all, bodies explicitly written upon). One need look no further than to the fact that it was not until May 2006 that Oklahoma lifted its 43-year ban on the practice of tattooing, or to South Carolina's and Rhode Island's prohibition of facial tattoos, or to state and local zoning ordinances for tattoo studios, to become aware that tattoos still inspire some degree of discomfort about who gets to do what to the visual character of the human body. In addition, when tattooed bodies are on display *as* tattooed bodies, they are the quintessential always already aesthetic image. With tattoos, there is no hiding or avoiding the fact that the human body is an aesthetic and visual event. Tattoos consequently function as both a site of, and as a productive metaphor for, visual analysis.

Through its focus on tattoo images, its particular articulation of agency, and its appreciation of history, *Tattoos in American Visual Culture* enters into the field of visual culture via performance and performativity. The mobilization of the term "performance" is meant to suggest the importance of evoking and attending to the embodied actions and movements remembered and represented in visual imagery. This understanding of performance infiltrates the book in many ways, but most obviously in the choice of the critical method of performance genealogy (explained at length near the end of this introduction). In the second case, the concept of performativity is mobilized to turn the model of embodied action toward the behavior of language and discourse itself. This notion, which I detail much more thoroughly in chapter one, facilitates questions about the activity and structures of language and normative discourse and how those activities and structures work with, through, in, and around visual imagery. Along these lines I argue throughout the book that images have the performative capacity to actively disrupt and destabilize social and cultural norms. Moreover, this vitality functions because of and through normalizing discourses and discursive strategies and not just in opposition or resistance to them. Images are more than merely texts that convey or contest meaning. Instead, visual images actively generate and participate in discourse. *Tattoos in American Visual Culture* insists, moreover, that the performative vitality of images depends both upon their immediate sociocultural context and, equally importantly, upon their connection to and revision of historical contexts and discourses. Through a genealogical analysis of the tattooed body on display in three sites of visual representation (tattoo contests, advertising, and photography), I show how visual images perform within and through the normative structures of class, gender, and exoticism to produce tensions that actively destabilize ideological discourse. Throughout its analysis of the tattooed body on/in display, this book consequently negotiates the themes of history, agency, performance, and visual representation.

Some of the key questions that circulate throughout this analysis are: What are the discursive strategies that circulate through and around tattooed bodies in the process of making and containing their meaning? What are the social and historical discourses and events that participate in meaning construction? How do visual representations of tattooed bodies participate in how those meanings are made? What, if anything, is it about particular contexts and forms of visual display that seem to have the capacity to produce instabilities in meaning construction?

METHOD AND ORGANIZATION

The general organization of the book moves from a theoretical account of agency through performance to three demonstrations of how agency emerges within specific discursive and image formations. From telling, in other words, to showing. The internal organization of each analytical chapter is inspired by the process of constructing a performance genealogy. Before providing an overview of the individual chapters, a brief detour into method is consequently warranted.

GENEALOGIES OF PERFORMANCE

Performance genealogy is a way to both attend to the importance of historical/cultural contexts to the construction and interpretation of meaning, and to the activity of discourse within and through those contexts. In his adaptation of Foucault's concept of genealogy, Joseph Roach suggests that genealogies of performance "document the historical transmission and dissemination of cultural practices through collective representations."[3] The genealogist assumes that current images and activities are present-day rehearsals of past events and discourses. My tattooed body walking down the street in a mid-sized southern U.S. city on a humid summer afternoon, for example, is a present-day descendant of numerous other tattooed bodies that have walked along the streets and pages of history. The aim of the genealogist then is to construct the image of the past rehearsed through the present event in order to show the process of cultural and discursive transformation. What are, in other words, the traces of the past found in and on my body as it walks down this particular street on this particular day? This is done by approaching the present cultural context as a complex puzzle of actions (kinesthetic memory), spaces (vortices of behavior), and texts whose scattered pieces contain fragments of images, words, and thoughts, which, once assembled, create a picture of cultural memory. The combination of these texts is designed to both provide a historical context as well as evidence of the different modes of communication that "have produced one another interactively over time."[4] In the end, performance genealogies actively and generatively interpret the history of the present through analysis and combination of textual fragments of cultural communication "under the rubric of performance."[5]

This method is designed to get at the process through which "culture reproduces and re-creates itself."[6] The genealogical approach to the analysis of the characteristics of the discursive practices involved in the construction of meaning is based, therefore, upon the notion that contemporary visual representations of the tattooed body are cultural reproductions (citations, reiterations) of earlier representations. This is a process that Roach calls surrogation. Surrogation occurs through culture's attempt to replace actual or perceived vacancies in social relations/cultural performances with alternatives. Cultural activities of the present, therefore, are both substitutions for and repetitions of real and imagined events.[7] Roach's example for this is the retirement of an individual from an academic post in a department at a college or university. That individual served not only the official functions of teaching courses and advising students but also served social functions. The person might have acted as a tension-reliever by telling jokes in meetings, for instance, or his/her office may have been a safe haven where others vented their frustrations without recrimination. These practical and social functions need to be filled in by the remaining people in the department.

Surrogation is also a selective and imaginatively imprecise operation of cultural change and transformation in that it involves processes of both remembering and forgetting. The surrogate will always fail to exactly "fill in" the cultural gaps. For instance, the members of the university department will never exactly replicate the social roles performed by their missing colleague. Cultural narratives and "myths of legitimacy" emerge in order to suture together those gaps.[8] These narratives fundamentally construct the "past" that the surrogate fills by recalling some aspects, changing and eliminating others. The colleague's therapeutic skills are either, for instance, completely forgotten, romanticized as irreplaceable, or demonized as meddlesome and happily dismissed. The activities of forgetting and remembering necessarily, therefore, involve the simultaneous alteration and construction of that which they forget and remember. Via these narratives and the discursive techniques that they employ, cultural and social meaning is created and the constant movement of cultural change is controlled. The tattooed bodies that we see on display in the early twenty-first century are, in other words, revisions and recitations of tattooed bodies that have been on display in the United States since at least the late nineteenth century. My tattooed body is also a surrogate. It is a distant descendent, reproducer, and creator of, as one example, the illustrated men and women who bared their tattoos on Barnum's stages. It is also a surrogate of female tattooed bodies imaginatively constructed through narratives of, as another example, sexual social deviance designed to control the cultural meaning of the tattooed body.[9]

Contemporary visual representations of the tattooed body are culturally evaluated based upon the social discourses used to interpret prior representations and by the revision of those discourses made possible by and through new social contexts. The contemporary image of the tattooed body cites the meanings of the past at the same time as it participates in creating and potentially altering that history. It is possible (though not guaranteed), for example, that the undergraduate student encounter with my white, female, professor tattooed body in the university fitness center, and the confusion that it incites due to the confrontation with various social discourses, may alter or distort the viewer's perception of me, of gender, of class, and even of the illustrated women performing on Barnum's stages. More importantly, the encounter may not only alter perspectives, it may actually *change* gender, class, and historical constructs. Contemporary visual representation of the tattooed body in the United States recalls *and revises*, consequently, the complicated history of the cultural display, categorization, interpretation, and evaluation of the tattooed body throughout the century.

The performance genealogist also approaches human bodies and social space as resources for cultural restoration and change. Bodies "kinesthetically imagine," through physical expressive techniques, cultural memory and context.[10] Roach indicates, for instance, that theatrical performances and the performance of everyday lives (as well as the memory of those performances located in representations) restore culture and history as well as actively expand and create them.[11] In other words, the actions of bodies in social space are resources of cultural memory and are understood according to their adherence to or violation of norms constructed through and by social discourse. Furthermore, kinesthetic imagination is also present in "systems of behavioral memory such as law and custom."[12] In order to attend to the relationship between discourse and bodily constructions, therefore, both the form of the body in representation and the memory of the body in action and representation must be addressed.

Additionally, space functions as a "site of memory."[13] "Vortices of behavior," according to Roach, "canalize specified needs, desires, and habits in order to reproduce them."[14] Space is a compact and heightened location where cultural meaning is produced and reproduced. Say, for example, I am walking my tattooed body down the mid-sized Southern city street on a warm March 17 during a St. Patrick's Day celebration. The street on this day in this city becomes different from the everyday and so people act differently in and on it. Moreover, the festival space also changes the ways in which people evaluate the actions and performances of those around them. In the context of the visual image, space becomes a matter of form. The spatial relations between and among form and content constitute the image's vortex. The power of place as the location for the reiteration and reproduction of cultural myths and narratives, therefore, becomes the activity of form within the static or graphic image. Genealogy thus activates the image by infusing form with the capacity to participate in the reproduction and production of meaning. Genealogies of performance consequently attend to the activity of context, history, space and bodies and so offer a method for attending to these issues as processes active within representational forms.

I adapt this method as a means to illustrate and map meaning as it circulates in and through visual representations of tattooed bodies. Genealogies of performance are uniquely suited to this purpose because they ask the cultural critic to produce as well as document cultural and historical discourses that accumulate in and through particular visual images and texts. The analytic procedure followed in each chapter is initiated with a description of the discourses and their performative structures. Each analysis consequently participates in the practice of interpreting and assigning meaning to tattooed bodies by "reading" visual images and identifying and evaluating the discourses that construct those meanings. This first step is critical because identifying how and what the image of the tattooed body means becomes the discursive structure and platform from which agency can then be potentially activated. Each chapter then points to moments where something different or disruptive either can or actually has happened within and through these meanings and discourses.

In addition to highlighting the structures of normative discourse, the process of analysis also functions to weave together a cultural history of the tattooed body on display. The genealogies constructed show the circulation of those discourses and images about and of tattooed bodies through the majority of the twentieth century in the United States. The chapters engage in dialogues between different decades and, when taken together as a whole, construct a loosely weaved chronology of U.S. tattooing in the twentieth century by adding new historical information in each chapter. Thus, history accumulates throughout the book in this process. The primary historical era discussed in chapter two, for instance, is the late nineteenth and early twentieth century. Chapter three then adds more of the mid-twentieth century, and chapter four concentrates initially on the tattoo renaissance in the 1990s.

Each chapter also selects a specific site of visual cultural production as its point of critical departure. The choice of these sites is based upon two issues: form of visual representation and the relationship between cultural producer and audience. In the first case, I desire to analyze pictures and images in the most extended sense by assuming that "a picture refers to the entire situation in which an image has made its appearance."[15] The picture is the socially, historically, and culturally framed visual event within which a discursively meaningfully image emerges. The phrase "visual image" is used throughout the book, therefore, to refer to particular moments of visual display that are a nexus for a specific constellation of social, historical, and cultural discourses. Pictures and images, therefore, arise not only in "static" forms such as the graphic arts (chapter three) and photography (chapter four), but also in patterns of human behavior or, in the terms of this book, performance (chapter two). In the second case, the goal is to analyze images produced by the tattoo community for the tattoo community (chapter two), produced by the mass media for a general audience (chapter three), and produced by the tattoo community for a general audience (chapter four). This emphasis on the producers and consumers is motivated by the understanding that production and consumption are critical factors in the process of representation and the circuit of culture.[16]

Chapter Organization

Chapter one describes the book's theoretical approach to agency. It argues that the agency of images can be productively located within the performative processes of reiteration and reproduction of discursive categories and structures. The chapter begins with a discussion of how the logic of representation limits agency to a model of social inscription pitted against individual expression. Next, the chapter illustrates how that model shapes tattoo scholarship and popular interpretations of tattooing. Through an articulation of theories of performance and performativity, I suggest that a necessary supplement to this model is found by looking within the repetitive and reproductive practices of discourse to see how images act. By drawing upon the philosophy of Gilles Deleuze and Felix Guattari to expand the idea of performance and performativity to images, I suggest an understanding of agency as the activity that occurs within specific discursive relations.

The following three analytic chapters enact the intersections of these ideas through performance genealogies of the image(s) of the tattooed body presented. Chapter two opens with the site of tattoo contests, the discourse of social and cultural class, and the discursive strategy of classification. The chapter begins with a discussion of class as a performative construct, drawing primarily from the ideas of Pierre Bourdieu. Then it focuses on describing both the performances at and the spaces of tattoo contests and conventions and how these performing bodies and convention spaces reenact the performances of illustrated men and women on sideshow and dime museum stages during the turn into the twentieth century. I show how that historical citation reproduces the discourse of classification as a strategy that normalizes the class evaluation of the tattooed body. I argue that the force of the normalizing discourse of class to contain and dictate the meaning of the (tattooed) body, however, is challenged through the actual and immanent movement of the tattooed body on stage. The resulting documentation of the tattoo competition, therefore, simultaneously shows the influence of history on interpretations of contemporary tattooed bodies on display as well as illustrates the ability of the contemporary context and performances to call historical authority and the construction of cultural memory into question. It is through the interplay between the classification discourse, the discursive history, and the present enactment that the moments of possible transformation ultimately occur. The chapter concludes by emphasizing the potential for agency that is always immanent in the display by showing how the explicitly tattooed body's actions overflow and subsequently dismantle the logic of class normalization even as they resist and recuperate it.

Chapter three moves from class to gender and from classification to fragmentation with its analysis of advertising campaigns that highlight tattooed body parts. The chapter starts with an explanation of gender performativity influenced primarily by Judith Butler's work. I then begin the analysis of gender and advertising by discussing a specific 1999 Tampax advertisement featuring a tattooed Rosie the Riveter, an advertisement that serves as an entryway into the discussion of gender and the tattooed body. The chapter demonstrates how the Tampax advertisement cites historical discourses of tattooed women on sideshow stages, of tattoos and sexuality in the mid-century, and of the history of the tattooed body as a marketing tool in advertising, particularly as it resurrects the image of the Marlboro Man. I argue that the strategy of body fragmentation is used to enforce normalized categories of feminine and masculine upon the body within the context of advertising. I also suggest that this fragmentation is a strategy resurrected through the history of women on display. Chapter three concludes by suggesting the power of recombination that is produced through the discourse of fragmentation.

Chapter four examines modern primitives, the discursive strategy of exoticism, and the tattooed body in illustrated nonfiction books. I argue for a performative conceptualization of exoticism supplemented by Homi Bhabha's notion of the "third space." The texts for this chapter are illustrated "coffee table" books with combinations of photographs and narrative captions, interviews, and analysis of the tattooed body as a Modern Primitive. I argue that the modern primitive tattooed body in these contexts is the product of a process of exoticism that both cites a history of exotic display of tattooed bodies as well as actively displaces the classed history discussed in chapter two. I suggest that it is through this simultaneous citation and displacement that the potential performative power of the modern primitive display is enacted.

Each of these case studies pursues a largely deconstructive agenda. Chapter five, therefore, puts together what has been taken apart. Following a brief summary of the possibilities for agency within visual representation discovered through the analysis of the tattooed body within the visual contexts of live performance, advertising, and photography and the discursive contexts of class, gender and exoticism, the conclusion turns to the potential for agency evoked through the process and product of the critical encounter. As chapter one suggests, the style of critical engagement is sometimes essential in provoking and/or evoking the agentic capacities within visual representation. The conclusion therefore concentrates on describing and encouraging alternative/experimental methods of critical encounter with the visual inspired by the analyses throughout the book as well as by the logics and languages of the tattoo community.

CHAPTER ONE

VITAL IMAGES: THE PERFORMANCE OF PERFORMATIVITY

THIS CHAPTER NEGOTIATES THE ISSUES OF VISUAL IMAGES AND MEANING through the concepts of representation, performativity, and agency in order to set the stage for critically conceiving agencies in visual representation. The first section describes a current conceptualization of agency in terms of the logic of visual representation, demonstrates how this notion of agency functions within tattoo discourse, and suggests the continuing need to explore alternative approaches. The second section provides the impetus for an alternative approach by discussing the conceptual value of performativity and performance in terms of the agency of language and the image. By drawing upon the work of Gilles Deleuze and Felix Guattari, the final section offers the description of that alternative approach to agency and suggests the importance of the nature of critical encounter to this alternative. Throughout the chapter, the discussion also sets the tone for the themes of history, representation, agency and performance laid out in the introduction and that run throughout the book.

REPRESENTATION, RESISTANCE, AND REPRODUCTION

The politics of representation are at the heart of contemporary cultural studies research and critical inquiry. Researchers in cultural studies have suggested that social power is produced, reproduced, and potentially resisted within social and historical discursive regimes cited and constructed through representation. Research in cultural studies has, therefore, undertaken the important task of exposing the power relations implicit in the process of representation. Stuart Hall, for example, famously argues that cultural studies "has to analyze certain things about the constitutive and political nature of representation itself."[1] The role of the critic is to focus on the actual process of representation and what it entails in order to expose the social and political possibilities and dangers in the practices of representation, be they verbal or visual. Representations, moreover, have no "natural" or inherent meaning separate from social and historical codes and contexts. One of the outcomes stemming from this approach to representation has been that representations theoretically become open to a variety of interpretations. Each text, according to Hebdige, for example, is polysemic or "seen to generate a potentially infinite range of meanings."[2] If, for instance, the reader begins to question the "natural" meaning of a sign and deconstructs the process of naturalization, then the "fact" of the representation becomes open to multiple interpretations. Textual analysis based upon this perspective concentrates on the process of meaning-construction in order to determine which meanings are privileged at any given time and whether or not the representation fulfills the possibility to contest the logic of normative and ideological reproduction.

These approaches to visual representation have been immensely important to the project of deconstructing and exposing ideological messages. In some cases, they are, however, limited in their capacity to generate a productive version of agency (social or political action) because they rely on notions of excess and exclusion. In the first case, the analyses value opening up the spaces of excess produced in the construction of meaning. "Representation," according to Peggy Phelan, for example, "follows two laws: it always conveys more than it intends; and it is never totalizing. The 'excess' meaning conveyed by a representation creates a supplement that makes multiple and resistant readings possible."[3] Individual and collective power over visual representations, therefore, is exerted only by interpreting them in such a way that thwarts or exposes the dominant social meaning. Alternative readings become the sole resources for agency or action when locked in this logic of representation. Individual subjects, in other words, can read visual objects by either exposing the visual object's representational politics and/or by offering resistant readings.

In the second case, exclusion is understood in terms of the "gaps and fissures" produced in the process of representation. Judith Butler argues, for example, that social norms produced in and through representations depend upon a repetitious process of meaning-construction that inherently excludes specific identity constructs and performances. Gender, for instance, is constituted in repetitive behaviors and representations that enact its standards. There is a photograph in William DeMichele's *The Illustrated Woman*, for example, that pictures "Jennifer."[4] The photograph is of Jennifer's makeup-free freckled face and shaved head. She has five tattoos on her scalp that form approximately two-inch-wide yellow, red, and blue stripes composed of a weave-like (perhaps Celtic-inspired) design that run from her front hairline to (presumably) the back of her head. She has multiple piercings (at least six) in each ear, three eyebrow piercings, four nose-rings, and her bottom lip is pierced. The flouting of feminine gender conventions in this photograph (no makeup, shaved head, excessive tattoos and piercings), in Butler's terms, enact standards of femininity because Jennifer's performance is read according to, and excluded by, those standards. Thus, "Jennifer's" being "not" typically feminine actually reproduces the category of femininity.

Butler argues that the process of repetition is, however, never exact and therefore "it is also by virtue of this reiteration that gaps and fissures are opened up as the constitutive instabilities in such constructions, as that which escapes or exceeds the norm, as that which cannot be wholly defined or fixed by the repetitive labor of that norm. The instability is the deconstituting possibility in the very process of repetition."[5] Butler's account highlights the discursive construction of subject positions as well as emphasizes the destabilizing force of repetition and citationality.[6] The gaps in Jennifer's photograph could be those aspects of the picture that trouble the easy interpretation of Jennifer as flouting feminine conventions. Her eyebrows seem manicured, her features are soft and pretty, her gaze innocent and open, almost inquisitive. The potential for deconstituting gender comes when we ask if her masculine and feminine signs can be reconciled or normalized within a potentially revised notion of female gender, or whether the contradiction excludes her from binary gender categories altogether. The specific manifestation of the force of destabilization, therefore, is that it either produces revised meanings or creates excluded *positions*. Either Jennifer's performance of femininity is normalized through an intersection between that category and "punk" (or some other subcultural type), or she is that which is excluded from gender altogether.

The focus on the *process* of repetition ends consequently once the meanings produced or subject-positions lost in reproduction and reiteration are identified (punk, anomalous other). Despite the generative conceptual work that Butler performs in emphasizing the continual process of performativity, the analytical practice pursued can tend to shift attention away from those performative processes by operationally defining agency as the ability to respond to or through the *products* of representation—be they meanings or visually and discursively excluded subjects/objects. Butler's analyses in *Bodies that Matter* also have a tendency to emphasize these abject positions, or the "constitutive outsides" of representation. The choices within the representational economy, therefore, seem to be either to resist the dominant interpretation by producing alternative readings or representations, or to focus on reading that which is outside of the representation in order to suggest its resistant possibilities.

These accounts of representation and the possibilities of agency through excess and resistance that I express through the work of Phelan and Butler are, of course, very familiar ones.[7] Despite the possible argument that (some) critiques of culture have moved passed this account of representation and resistance into more productive territory, Stephen Crofts Wiley has recently argued that much of cultural studies research has retained this "epistemological problem of representation."[8] That is, when it comes to figuring out how to deal with the power of visual images to construct, reveal, and participate in changing cultural norms and ideologies, the choices remain relatively limited to strategies of interpretation. In fact, as Hans Ulrich Gumbrecht suggests, the interpretation of meaning remains the "exclusive core practice" of humanities research.[9] It is partly this dedication to hermeneutics and its emphasis on "individuated psychological interpretations and tastes" that has lead Lawrence Grossberg to call for forms of critical engagement that go "somewhere else."[10]

I find it valuable to rehearse the specific contours of this problem within the context of visual representation because they continue to resonate both in the way agency is understood in general, and in particular on how agency is conceptualized within interpretations of tattoos in both the popular and academic press. As the following discussion will demonstrate, in both cases, agency is the subject's ability to act in response to the meanings and messages of images and discourse. That is, individual or collective actors can either create alternative images that produce resistant meanings or can produce alternative interpretations of images in order to assert social power.[11] The problem with this articulation of agency is not that it is wrong. The political and social efficacy of producing resistant readings and alternative interpretations is not in question here. Rather, the limitation with this view is that it does not embrace the possibilities for change and transformation that operate *within* the process of meaning making. The reliance on interpretation and on locating the existence of excess and exclusion has resulted in neglecting the potential power located in the actual process of reiteration and representational reproduction: the power, in other words, within performative processes of meaning making. Before exploring those possibilities more fully, however, it is now necessary to explore the ways in which the epistemology of representation has shaped tattoo scholarship. This narration demonstrates exactly how the logic of representation seriously constrains the possible resources for social action of the tattooed body and of visual images.

The Subject in/of Tattooing

The primary theme in this narrative of agency and the tattooed body is the interpretation of the tattoo as a visual mark of deviance cast either by an individual subject upon the self or cast upon a deviant subject by an institution. Within this discourse, two themes emerge. First is the tattoo's function of making the body into a visual representational text. Second is conceptualizing tattoo agency in terms of either resisting inscriptions of social norms of deviance or creating representations through individual acts of resistant expression.[12] This story of deviance begins with historical accounts of tattooing. Tattooing prisoners has been one practice highlighted to demonstrate the history of tattooing's deviance, as well as the force of the institutional power of the state in creating that meaning. Gustafson argues, for example, that the Roman practice of tattooing as punishment, and the use of the term stigma to refer to this practice, was an effort to discipline the body and to permanently mark its deviance:

> [T]hose in power were well aware that the body can function as a permanently running advertisement of one's guilt and subjugation. Given the heavy yoke of a tattoo, those released from their sentences and allowed to return home (as sometimes happened, for various reasons, including old age and infirmity, or as the result of a general amnesty at the beginning of a new reign, or other imperial indulgences) could never completely resume normal life. Unless, of course, they could get rid of those "indelible" marks.[13]

The status of the body as a visual text is obviously imperative to this account of tattooing as a sign marking and subsequently advertising deviance. Alternative interpretations based in a Christian ideology of the time also rely on the body as text metaphor, but this time it is a positive resource. Gustafson suggests, for example, that tattooing also served as a badge of Christian honor due to a "paradigm shift" marked by the Apostle Paul's reference to being tattooed with the marks of Jesus.[14] The practice of tattooing, according to Gustafson, was both utilized as state institutional discipline and as an individual expression of spirituality. In other words, in Gustafson's analysis, tattoos function as both oppressive social inscriptions of deviant norms and resources for individual representational practices.

MacQuarrie also takes up the theme of simultaneous stigma and honor in his analysis of the Celtic tradition of tattooing. MacQuarrie argues that references to tattoos in the classical and medieval world associate the practice with "both Christianity and with demons, with both civilization and the forces of disorder."[15] Tattoos are read as "signs of deviance or self-discipline" depending on whether the body tattooed is from a "civilized" or "savage" culture.[16] In MacQuarrie's historical account, the meaning of the tattoo depends both on cultural norms and on whether the mark is chosen (an individual act of resistance or adherence) or imposed (a social stigma inscribed). The historical use of the tattoo as a metaphor for social inscription and individual expression is clear in this account. This explanation also rehearses the pattern of theorizing tattoo agency in terms of a dialectic between oppressive social inscriptions and individual acts of expression and resistance.

Interpretations of nineteenth century tattooing practices in Europe and European colonies continue to swing the pendulum back and forth between the social and individual and between resistance and expression or domination. For instance, Maxwell-Stuart and Duffield argue that Australian convicts used the imagery in their tattoos to express resistance to the disciplinary "shaping" of their bodies through means such as surveillance and torture.[17] Schrader's analysis of the practice of convict tattooing in Russia and the Soviet Union similarly suggests that inmates appropriated the process of bodily inscription in order to "assert ownership over their bodies."[18] Despite the efforts of convicts to recode the meaning of tattoos, however, Caplan's analysis suggests that nineteenth-century European tattooing remained largely understood through the discourse of criminology.[19] The act of tattooing criminals, the continual cultural association between criminal bodies and tattooing, and the simultaneous belief that tattoos are the expression of the self, all eventually lead to the belief that tattooing itself is a sign of a criminal mind. This rhetoric even influenced Victorian British accounts, despite a tendency there to issue environmental and contextual (and importantly classed) interpretations of tattooing.[20]

Sociological investigations into group identification and criminal and social deviance also replicate the representational logic constraining agency to resistance and expression. A variety of scholars, for instance, have categorized the different groups of people associated with tattooing. Bikers, convicts, sailors, punks, and juvenile delinquents are identity categories associated with tattooing during the mid-twentieth century prior to 1960.[21] The primary explanation for the popularity of tattooing among these groups argues that tattoos symbolize subcultural group affiliation. The specific deviant meaning of the tattoo changes slightly, but the structure of that meaning reproduces the tension between social and individual inscription and expression. The visual sign imposed upon the criminal body, which subsequently became an indication of a criminal or deviant mind, is now a badge of deviant social class affiliation or an individual expression of identification with deviant subcultural identity formations.

Explanations of the meaning of early to mid-twentieth-century military tattooing practices also reproduce the dialectic of social inscription versus individual expression. Alan Govenar suggests, for instance, that tattoos "fortified the masculine egos of the wearers or vented the frustrations and anxieties of war."[22] In short, the visual body text either reinforces the social norm of gender or expresses the individual experience of trauma. Military and governmental responses to tattooing, moreover, became a popular demonstration of disciplinary control when "both the army and the navy publicly stated their disapproval of tattooing" and, after the war, "health authorities, in a combined effort with the police and courts, began to inspect tattooing practices more closely."[23] The force of these institutional and social discourses was strong as many servicemen began to recognize the negative social value of their tattoos. They subsequently sought to have the tattoos removed or, alternately, explained the tattoo away as a result of drunken decision making.[24] Again, the tension in assigning meaning to tattooing, both socially and theoretically, is between the inscription of disciplinary power upon the body text, and the mark as an expression of some kind of deviant and/or resistant interiority.[25]

It is not only in the historical, popular, and sociological accounts of tattooing, however, that the discourse of deviance and the inscription versus expression dialectic dominates. Psychological accounts of tattooing also reproduce this representational logic through their rhetoric of deviance.[26] In these studies, tattoos are treated as symptoms of an internal self—as, according to Sullivan, "the expression (whether unconscious or conscious, it makes little difference) of a sign designating the truth of an innate 'I.'"[27] Sullivan calls these types of analyses of tattoos the acts of "dermal diagnosticians." She persuasively argues that all of this literature, though nuanced in particular and specific ways, depends on the underlying assumption that tattoos can be read as expressions of individual personalities. The vulgar and tasteless tattoo, for instance, becomes an expression of the vulgar and tasteless self.

Thus far, the majority of the research I have discussed in demonstrating the relation between social inscription and individual expression in the making of the deviant tattooed body has been interested in tattooing prior to the late twentieth century. This is because there was a shift roughly around the 1960s in the content (if not the structure) of meaning making surrounding tattooing that continues to persist into the twenty-first century. After 1960, a generation of tattoo clients and artists emerged resulting in what is now often called the "tattoo renaissance."[28] The renaissance is characterized by a larger number of middle-class clients, artists with a higher level of formal education, and an increasing appreciation for the artistry of tattooing by both artists and clients. Margo DeMello argues in *Bodies of Inscription* that the renaissance was occasioned during the 1960s and 1970s by the rhetoric produced from movements such as New Age, ecology, women, men, self-help, and Goddess religion.[29] As a result of this new rhetoric (as DeMello suggests), investigations into post-1960s trends in tattooing have been dedicated to collecting personal narratives in order to understand why people opt for tattoos and what tattoos mean to individuals in terms of their identity within particular historical, social, and cultural contexts. In short, the difference between studying pre-1960 and post-1960 tattooed bodies is a shift from largely historical, cultural anthropological, and psychological modes of inquiry to more (though not exclusively so) ethnographic, sociological, and critical cultural approaches. In terms of my tour through this scholarship, this shift does not suggest a change in the importance of meaning, but merely an adjustment in who gets to claim authorship (agency) over meaning. As a way to close this discussion about the conceptualization of agency as resistance within a binary between social inscription and individual expression (as operationalized in tattoo interpretations), I turn to two final recent examples, one academic and one popular.

The theme of socially inscribed discipline upon the tattooed body and the potentially resistant act of self-inscription is taken up by Victoria Pitts's account of contemporary tattooing practices in *In The Flesh: The Cultural Politics of Body Modification*. Pitts speaks primarily of the actual practices of body modification and not about the visual representations of the embodied results of those practices. Pitts's commentary about the means through which bodies are inscribed with meaning does, however, resonate with my articulation of the mechanism through which meaning operates in visual representation. Pitts's sophisticated account of late twentieth century body-modification culture and practices questions the idea of the self using representational practices to subvert or resist power because "self invention is an ideology."[30] Skeptical of the notion of self-narration as an ideologically unmarked practice, Pitts insists upon the primary force of social discourse(s):

> When bodies are understood as social and political—as inscribed by and lived within power relations—anomalous body modifications do not appear as inherently unnatural or pathological, but they also don't illustrate that individuals can freely or limitlessly shape their own bodies and identities. Rather, body projects suggest how individuals and groups negotiate the relationships between identity, culture, and their own bodies.[31]

While Pitts' point is well-taken that the self is a suspect category when it appears as isolated and separate from social discourses, her argument assumes that bodies are either imposed upon (inscribed) by power or operate within power but do not produce power. Bodies, in her terms, either "express or reflect social control."[32] Pitts's argument that social discourse is the dominant force that individuals and groups must negotiate reinforces the idea that the body/subject is inscribed at the will of some power external to it. This approach grants agency to ubiquitous "power" and presents individual acts of resistance as made available only in and through the meanings of the discourses that they resist. The end result is a predictable one in the terrain of critical cultural criticism: both/and. In Pitts's case, radical body modification practices are *both* subversive *and* not subversive. In applying that logic to visual representation in general, representations are *both* resistant *and* ideologically reproductive. This both/and position is paralyzing for political agency (not to mention entirely too predictable) because it resolves the conflicts within these representations by providing a (un)happy compromise: that the two positions can coexist but not have to interact. The dominant and subversive meanings reach a détente of sorts that diffuses the relational tension at the border. They can agree to disagree.

For the most part, then, tattoos in the popular and academic press remain primarily understood as symbols inscribed upon bodies by either oppressive social institutions or by autonomous individuals striving for self-expression. One final popular cultural example of the prevalence of this discourse in the understanding of the social and cultural role of U.S. tattooing drives this point home. *Miami Ink* is a reality TV show aired on the The Learning Channel (TLC)—a Discovery Communications television network. The basic structure of the show is designed around the lives of tattoo artists working in a Miami-based studio and the stories of their clients. A typical episode will include a tale of someone getting a tattoo beginning from the moment that they enter the studio and consult with one of the artists, progressing through the artist's rendering of the image, and ending with the actual tattoo application. Despite differences in plot and theme, the common narrative explaining nearly every tattoo focuses on the individual's choice to represent some event, person, emotion, or aspect of their identity through a specific design they have chosen to have inscribed on their body. Moreover, oftentimes there are moments embedded within this narrative that report reactions from friends and relatives to the choice to become tattooed that range from shock, to dismay, to incredulity. Despite (or, in some cases, fueled by) objections, however, the individual persists with their desire and, in the highly public space of cable television, gets inked. The history of the tattoo as a sign of individual expression and countercultural resistance clearly circulates both in the stories of the individuals and the editing choices of the show's director and producers. The point is not that the people represented on the show are lying or being misrepresented, but that the understanding of the tattoo as an individually chosen symbol that has a specific and controllable (albeit private) meaning is locked within the logic of expression.

Additionally, the continuing association between tattooing and somewhat deviant expression is maintained both by the narratives of familial objections as well as by some of the characterizations of the clients and artists. For instance, the video biography of tattoo artist Chris Nuñez highlights the fact that he was twice deported from Brazil and is known for being "shipwrecked" alone in bars late at night.[33] The popularity of the television show assures that this particular understanding of tattooing will continue to circulate. These deviant characterizations are also not total misrepresentations. In her participant observation of the "elite tattoo realm" in the 1990s and the early part of the following decade, Katherine Irwin discovered that elite tattoo collectors and artists actually do embrace "deviance" as a mechanism to separate themselves from the popularization of tattooing.[34] By choosing tattoos that violate appearance standards, that represent "fringe" themes, and by celebrating continuing associations between tattoos and deviant social groups, Irwin argues that elite tattoo community members celebrate deviance as an identity.[35] Moreover, this understanding permeates the meaning of images of tattooed bodies. That is, pictures of people who have tattoos tend to be interpreted using this same logic.

This extended narration of the deviant meaning of the visual image of the tattooed body has hopefully emphatically illustrated the continued force of a representational logic that limits agency to social inscription and individual resistance. As the popular tattoo discourse illustrates, the choices for agency have been and continue to be generally constrained. Once the tattoo enters into public circulation as an image, moreover, these limitations extend into interpretations of the tattoo as an image. The tattoo image is both resistant to and reproductive of (gender, class, "normal") ideology. The contribution of this book is to attend to the specific discursive practices of meaning making in order to suggest an alternative route to agency. The vitality of images articulated throughout this book lies within the "fertility of reproduction" and not solely in resistance to meaning.[36] This shift requires a turn to performance, a revised orientation toward agency, and a consideration of history.

The Performance of Visual Performativity

In order to focus on the activity that occurs within meaning-making processes, and to use that as a foundation to conceptualize agency, I find myself turning to the concept of performativity. Performativity accentuates the importance of focusing on the process of meaning instead of on its discursive or representational products. In short, performativity facilitates the question: what are the forms and styles of discursive performances? In the following section, I will first briefly rehearse definitions of performativity and performance and then discuss the advantages of using these concepts in understanding visual agency.

Performativity is a useful term for getting at the activity of the reproduction of meaning. However, the term is often used in a variety of ways that confuse the conceptual distinctions between performance and performativity. For instance, one could refer to a visual image as being "performative" and mean that the image evokes or mimics the qualities of live performance or performs specific communicative functions.[37] One could also talk about the artistic activity of creating visual imagery as performative.[38] That is, the work of the artist is a heightened aesthetic activity.[39] The lineage of the term I am interested in is linguistic and describes the citational and repetitious practices of meaning making within representation that Butler, for example, identifies and theorizes. This lineage has often been traced through the work of J.L. Austin and Jacques Derrida (to name a few key figures in the stages of the development of the term).

J.L. Austin originally proposed the term "performative" as a means to distinguish between utterances that state facts and those that perform acts.[40] Performatives are acts with specific intentions located within a concrete context. For instance, the utterance "I do" becomes an action (marriage) within the context of a Christian wedding ceremony. The statement, "I am tattooed," on the other hand, merely names a condition constituted by a past performance. In "Signature Event Context," Derrida extends Austin's work with the argument that the source of power of performative utterances is located in their citation of authority.[41] In other words, "I do" as an utterance is not culturally productive without reiterating and citing a series of actions and utterances that came prior to its iteration and provide it with authority (the history of marriage ceremonies and legal and religious discourse, for example). Judith Butler, citing Derrida, further suggests that performativity is the citational process where discourse creates what it regulates.[42] When the current utterance (I do) refers to prior discourse (i.e., religious and legal), that utterance both performs the act of marriage and controls what that act is and means. Furthermore, the process of citation or reiteration is where discourse receives its force, its ability to act or produce that which it names.[43] In other words, the phrase "I do" is not felicitous (successful) if it does not cite the discourses surrounding wedding vows (if the phrase is uttered spontaneously in a pub at two o'clock in the morning, for example). Butler explains that citationality explodes the notion that performativity is a singular act or utterance because performativity is "always a reiteration of a norm or set of norms" that "conceals or dissimulates the conventions of which it is a repetition."[44] Norms are reproduced, naturalized, and altered through the repetition that constitutes their construction.

As this description indicates, performativity can be and has been used to concentrate on the normative products and naturalized meanings produced through and excluded by discursive practices. What is also true is that its description of the linguistic process provides a location for the analysis of the characteristics and structures of the discursive practices that produce meaning—not only in terms of what they exclude, but also in terms of how they *act*. Focusing of the process of performativity allows one to ask how, in other words, reiteration, citationality, and reproduction actually and specifically function. What are the characteristics of the specific performances of performativity?

To ask this question requires an understanding of performance itself and how it can be mobilized as a critical resource. In *Performance Studies: An Introduction*, Richard Schechner draws the useful distinction between the study of self-evident performances (e.g., dance, theater) and the study of texts and social events *as* performance. In the former case, the critic describes, interprets, and evaluates theatrical and performance art events for their aesthetic, cultural, political, and technical value. In the latter case, critics approach a variety of cultural artifacts through the metaphor of performance. For instance, sociologists influenced by Erving Goffman have analyzed everyday life as if it were a theatrical performance with characters and props, stages, and set pieces.[45] Performance in both these senses tends to be a "public display of technical skill."[46] Performance analysis describes and evaluates, therefore, the behaviors and aesthetic products of either professional performers or everyday social and cultural activities. Another version of the performance metaphor produces cultural criticism more concerned with the ritual and cultural functions of the performances themselves—how these behaviors enact and produce community or identity, for instance.[47]

Approaching culture by means of performance also opens the door to a more metonymic orientation, however.[48] One of the premier characteristics of performance is, of course, the actor's capacity to "stand-in" for a character, and the ability for the theatrical or ritual event to "stand-in" for "real" life. The representational logics of mimesis and metaphor can be productively supplemented by the performance logic of metonymy or substitution. The image and the discourse no longer become understood according to principles of performance, or even for their cultural effects or function, but for how they move or act. Replacing representation (mimesis, metaphor) with performance animates the body of representation by highlighting its performative function as opposed to is representational effects. This approach consequently facilitates a change from the "both narrowly idealistic (intentional) and deterministic" agency model, to a model of agency "as embodied action, as that which is generated in and as performance."[49] That is, by substituting human behavior for linguistic behavior, questions about the repetitive and "restored behaviors" of *language* become possible to ask.[50] Going beyond merely describing the performative sequence or pattern of repetition and reproduction, one can attend to its specific gestures and movement and their generative (or reductive) potential.

This performance approach to the performativity of language consequently constitutes the platform for a query into the agency of the forms and styles of movement within citational and reproductive processes. What becomes interesting are the small moments of potential or possible instability and change that occur *as* and *how* meaning is being reproduced. Rather than concentrating on the spectacular aspects of visual events, therefore, this perspective privileges the potentially mundane.[51] As a result, this perspective begins to address the "reproductive potency or fertility" of an image.[52]

Performing Images

Images, from this point of view, are living things. The particular form of life is not, however, analogous to an acting individual agent or subject. Instead, the images' life is rather more similar to the potential for life imagined in, for example, religious icons or, seen another way, the reproducibility of a biological virus. In the first case, images live because of the human tendency to attribute life. There is a propensity (both historical and cultural) to approach images "as-if" they have immanent power and life. Think about, for example, the role of material objects and images in certain religious rituals. The now consumerized, secularized, filmed, and televised practice of putting pins in a "voodoo" doll to effect pain in the person it models, for instance, suggests the attribution of life-force to objects and images. Images also live, however, because of their reproducibility.[53] Like the unpopular virus, which can only reproduce itself by joining with or passing through another cell, images are replicated and modified and incessantly circulated. Images, in W.J.T. Mitchell's appropriation of advertising vernacular, "have legs": "that is, they seem to have a surprising capacity to generate new directions and surprising twists in an ad campaign."[54] The critical position Mitchell advocates as a result of this approach to the life of images encourages, consequently, "a subtle dislocation of the target of interpretation" that endorses the "constitutive fiction" of pictures as "quasi-agents."[55]

The implications of this dislocation for the performance of reproduction and its analysis are threefold. Initially, it is an invitation to continue with the process of interpreting images by locating moments of representational excess. Dislocating interpretation does not mean refusing it altogether. In fact, much to the contrary this approach *requires* that interpretations (both resistant and reproductive) continue to be made in order to identify the strategies and structures of reproduction. Producing resistant and oppositional readings remains an important interpretive endeavor. There are instances throughout the book where I identify points of representational excess and escape and where I produce resistant readings. Secondly, as illustrated in the importance of performativity to this analysis, this approach supplements those resistant readings with an attention to the fertile and potentially generative process of reproduction itself. The dislocated target becomes the meaning-making process. Although Mitchell follows the legs of the image to see where they lead, the relocation of agency into the performance of reproduction is more interested in how they get there.

Most importantly, the emphasis on the life of the image encourages the turn to performance. The notion of the visual image as a "quasi-agent" asks us to think about how images function, how they act. Mitchell's approach to their agency, as indicated very directly in the title of his book, places the image in a subject position by asking *What Do Pictures Want*? For Mitchell, the picture or image becomes the agent/actor either through the interpretive effects it produces or through the attribution of speech acts to images. For instance, near the end of the book, Mitchell directly indicates, "I propose what I hope is a more nuanced and balanced approach located in the equivocation between the visual image as instrument and agency: the image as a tool for manipulation on the one hand, and as an apparently autonomous source of its own purposes and meanings on the other."[56] This account of the agency of images, however, firmly locks the image within the subject/object resistance/reproduction economy by attributing an active or subject status to the image. It is either used as a medium of communication, or it is a social actor.

As ought to have become clear by now, the addition to agency that I am offering instead locates power within processes of discursive reproduction and within the performance of performativity—that is, to look at how discourse acts as it is reproducing itself. What are its gestures and movements? What are the shapes and forms? Moreover, how does that process operate at the site of the visual image? How does representation happen in and within images? To conceptualize agency as the performance of performativity requires an entry into the meaning-making process in order to encounter the (sometimes mundane) moments of activity that propel images to move in potentially unanticipated directions.

AGENCY AND TERRITORIALIZATION

Attending to the potential power located within the processes of representation requires a reorientation toward the function of a social agent. The interpretive move of understanding the production and reproduction of meaning is supplemented by an attention to the ways that visual representation acts *through* meaning making. Attention is directed *within* the processes of representation and signification as they performatively reproduce and totalize. *Agency*, in these terms, *is the enactment of the potential for disruption, destabilization, transformation, and change that is produced in and through reiterative, reproductive, and citational visual and discursive practices of meaning making and meaning attribution*. This focus on agency through performativity thus encourages a type of criticism that does not "dream of getting beyond images, beyond representation."[57] Instead of "smashing" the images like piñatas in order to expose their hidden bounty of ideological oppression, the critic's job is to attend to the patterns of resonation, to follow the movements of the images' legs.

As the prior discussion suggests, performativity and performance are concepts that help to facilitate this focus on the reproduction of meaning as an activity of the body of representation. What the following discussion adds is a critical orientation that makes it further possible to attend to the specificity of the movements of these bodies of representation—a way to, in other words, concentrate on how repetition and citationality specifically perform. What is offered to get at these specific operations within meaning-making processes (the performance of performativity) are a terminology and a conceptual framework that facilitate a type of precise engagement at the level of discursive reproduction. Moreover, for the purposes of image analysis, this terminology takes into consideration the mobility and life force of the image. It is a critical orientation, in short, that embraces the "as-if" of the image.[58] This might initially seem to be a problematically utopian or, at least, a fictional critical position. It is fine, one might argue, to imagine that images have life for the purpose of entertainment or ritual, but to argue for political and social agency from the point of view of the living image might threaten to trivialize the issue. It could make cultural critics into naïve caricatures screaming "It's alive!" at portraits and photographs. One may argue, furthermore, that this theoretical posture neglects and belittles the harsh material realities of daily lives and the possible oppression authorized through (sexist and racist, for instance) visual images.

Despite that criticism, however, the "as-if" posture can and has produced "real" material effects. Moreover, this is particularly the case when it comes to the visual image. Mitchell's discussion of the second commandment's prohibition of the creation of images, for example, illustrates the real effects of the "as-if" in terms of idolatry. Mitchell suggests that early followers of Christianity took very seriously the idea that worshipping or creating a false image was much more than a symbolic blasphemy, it was an actual construction of another, albeit false, god. The destruction of the image became both the desecration of the false symbol as well as an annihilation of the untrue god *itself*. The idol (image or statue) was both the symbolic and actual body of the deity and thus its destruction was its death.

Despite the late-twentieth century postmodern critiques of the status of image as, for example, pure and infinitely interpretable surface without any concrete material referent, the connection between the visual image and "real" life continues, moreover, to have "real" effects.[59] As an example, Mitchell points to how aghast his colleague's students were when they were asked to cut out the eyes in a photograph of a family member. Similarly, a friend and former colleague of mine used to have his students draw three pictures. The first image was that of their most hated food, the second of an ugly bug, and the third of their mother. Then, in that order, he would ask them to put the paper on the floor and step on it. Invariably, the students were confused about stepping on the food, reveled in stomping on bugs, and balked at the idea of tramping on their mothers. On a more global, ideologically loaded, and humanly devastating level, as I am writing, there continue to be violent protests around the world in response to the publication of cartoons in a Danish newspaper that satirically picture the prophet Muhammad as, among other things, a terrorist with a bomb planted in his turban. Although much of the editorializing surrounding these events tends to be about the relation between free speech and hate speech, this is, above all else, an image event.[60] Whether or not one "believes" in the actual "life" of images is not the point. At issue is the force of that idea to have effects in terms of relations to and among images, to have effects on how images and discourse culturally and socially act.

In order to construct this version of agency as a performance of discursive reproduction enacted through and by the life of images, I draw from the ultimate philosophers of the "as-if," Gilles Deleuze and Felix Guattari. Deleuze and Guattari are important because their work creates "styles of thinking" that, among other things, "compel us to believe that to aestheticize is to politicize."[61] Deleuze and Guattari's philosophy does not, therefore, trivialize literature and art or politics. Rather, I understand their philosophy to be deeply concerned with articulating and producing possible futures by encouraging alternate modes of engaging the world. Deleuze and Guattari's description of assemblage and the process of territorialization thus provides the final conceptual foundation for my understanding of the performative agency of/and images. These concepts create the possibility of agency as being the active force of disruption and transformation occurring within discursive and ideological containment or capture—that is, agency within discourse and agency within images. Throughout the following description I draw upon the value of the tattoo as a metaphor that I alluded to in the Introduction.

Assemblage

The term assemblage describes patterns of organization (stratification) and processes of change (transformation) within the social (human expression and action) and natural worlds as a network of constantly moving relations. Within this network, there are instances when social, natural, and/or discursive spaces emerge that are ordered, closed, homogeneous, regularized, and highly structured and stratified with clear patterns of linear organization. Deleuze and Guattari illustrate these types of "striated" spaces through terms such as "stubborn geometry" and "geological foundations" and by descriptions of, for example, the military or the "war-machine."[62] Striated social space is something like, for example, forms of disciplinary power (Foucault) or ideological apparatuses (Althusser). There are also instances within this process where spaces emerge that are more supple, open, heterogeneous, variable, loosely structured, and smooth, with vague mutational configurations that are multidirectional. Deleuze and Guattari illustrate these types of "smooth" spaces through terms such as "nomad" and "fusion" and associate them with aerial acrobatics and the desert. This type of social or discursive space could be compared, for instance, to the motions of the sea, or the immersion into a space that folds over itself in a condition of constant overcoming.

The network of the assemblage consequently functions along two axes. The first axis describes moments where the network is organized into a "striated" territory (a relatively concrete meaning or image, for example). The second axis is "constituted by lines of deterritorialization" that shoot through, cut across, and open up the territorialized assemblage to make it become otherwise.[63] These "lines of flight" can transform the assemblage through, for instance, a process of mutation by infecting the striated territory like a virus. The lines can also transform the territory by stretching and extending it out in other directions (like stretching pizza dough with your fingers) or by moving it completely. Due to the complex associations among the smooth and striated, the relation between these axes is not consequently one of simple opposition:

> No sooner do we note a simple opposition between two kinds of space than we must indicate a much more complex difference by virtue of which the successive terms of the opposition fail to coincide entirely. And no sooner have we done that than we must remind ourselves that the two spaces in fact exist only in mixture: smooth space is constantly being translated, transversed into striated space; striated space is constantly being reversed, returned to a smooth space. In the first case, one organizes even the desert; in the second, the desert gains and grows; and the two can happen simultaneously.[64]

Within any territorialized or striated space, therefore, there is both patterned movement (linear, cause-effect) as well as the simultaneous potential for disruption or mutation.

In tattoo terms, striated space can be thought as the final version of the tattoo rendered upon the skin of the client. It is a frozen and structured image permanently inscribed. The tattoo image is, in these terms, a territory. Of course, tattoos are also written upon the largest living organ of the human body, an organ that is in a constant state of change and transformation as cells are reproduced and shed. Throughout this organ blood flows through tiny tubes of porous boundaries. Skin is also the organ that mediates between the human body and its physical environment. The impact of the force of this mediating role is inscribed in the skin in the form of cuts, wrinkles, and scrapes. It is also apparent in changes in texture and pigmentation. All of this skin-action works upon the tattoo. The colors fade, the image blurs, and the lines lose their integrity. As you will read in the next chapter, I got my first tattoo when I was 21. Instead of following the good advice of the artist and keeping it out of the sun and applying regular doses of a hypoallergenic and water-based lubricant, I immediately removed the bandage and sported a tattoo-revealing tank top through the summer afternoons. After years of subsequent exposure and aging and by the time I was in my late twenties, the once proud unicorn's head had mutated and deteriorated into a nearly unrecognizable morass of fading blue lines. One observer asked (not unkindly), "Why do you have an old man tattooed on your shoulder?" The constant movement of the skin assemblage acts on and through the seemingly static territory of the tattoo, thus illustrating the moving relations within territorialized space.

Additionally, the movement in territories *trans*lates and *trans*verses, rather than *inter*acts and *inter*sects. Lines and temporal space move beyond and through each other instead of crossing at designated points. Moreover, the movement within the territory can be cut through by a line of flight as well as have its own rhythm (repetition, reiteration). Striated space occurs as a process of formation and reversal. Like the leaves that fall from the trees in my front yard every autumn, striated space emerges into a form, only to be redistributed by a force such as the wind, my husband's rake, or the slow decomposing force of air, moisture, and time. I like alternately to think of this as a massive, three-dimensional, multilevel virtual pinball game where the routes (performative processes) are constantly changing and the balls (meanings or codes) can dematerialize and rematerialize, speed up, slow down, and even change shapes or be momentarily in suspended animation. Or, of course, we have the skin with is multiple and transecting layers cut through and across with veins and vessels, follicles and glands.

Sedimented meaning and material form are the crystallized (striated) moments of this process of arrangement: instances, in other words, where the speed slows intensely down so as to suggest a halt or a stable structure (to suggest a point of intersection). This is one reason why, I suppose, the skin seems so unlike an organ. It appears to be a static, unmoving layer of material. To shift examples, relatively stable-appearing meanings and interpretations are formed out of discourse and discursive strategies at the seemingly static location of a text or image. These moments are territorializations or the *making* of a territory out of (a moment in) the assemblage. Another way of putting this is the making of a tracing out of a map (when the map is not already a tracing but a multidimensional, multilayered network of lines in/of motion); as another example, it is the act of interpreting and evaluating a behavior according to a predetermined set of diagnostic categories.[65] In tattoo terms, we can think of this process in two ways. The first is through the tattoo itself. The tattoo as a work of art makes a relatively stable image out of the fluid movement of ink and blood, cells and dermal layers. On the other hand, the act of diagnosing the sign of the tattoo as a mark of deviance is also an act of territorialization. Territorializing effects happen through a variety of mechanisms (or, in Deleuze and Guattari parlance, machines).[66] In the case of the construction of the tattoo as a sign of psychological deviance, the process of territorialization occurs through the striated mechanisms of state or institutional biopower (the prison).[67] It also occurs through the discourse of psychology and the premise of the body as a site of either conscious or unconscious self-expression.[68] These discourses function as interpretive forms of territorialization that momentarily freeze the movement of part of the assemblage circulating within and through tattooed bodies and their images. They are tracings or stop operations when/where meanings are assigned.[69] The tattoo, then, is a moving aesthetic territory where the territorializing practices of meaning making occur.

In the process of describing the assemblage for the purpose of analyzing cultural expression, some critics tend to emphasize either points of intersection between lines of flight and territories, or the capacity for the territory to change by lines of flight. For instance, an assemblage might be a "site at which a discursive formation intersects with material practices,"[70] or maybe it is "a concept dealing with the play of contingency and structure, organization and change."[71] The former articulation is useful because it helps to visualize the assemblage as a place where discourse happens. It also, unfortunately, evacuates the inherent motion of assembly by limiting it to points of intersection between discourse and materiality. The latter description embraces the movement of the term and therefore provides a sense of place as that which is not a point of intersection in space, but rather a network or system of movement. Instead of a concrete site where ontologically distinct objects of discourse and materiality intersect, then, the assemblage is the contingent and changing activity of/through discourse and materiality that can occasionally produce moments of organization and structure. The assemblage, then, is not a locatable place where, in this case, discourse happens. Rather, the assemblage is an occasionally structured, multi-dimensional, and variable rhizomatic network of, for example, discursive and material "lines and measurable speeds."[72] This articulation accounts for the "trans" movement of the assemblage but highlights the positivity of lines of flight "opening" up "possibility" and concurrently emphasizes the negativity of the closing down operations of territorialization.[73]

Deleuze and Guattari's insistence that the smooth and striated are not oppositional spaces, however, suggests that there is potential for change *within* tracings and territorializations. Their choice of the term *agencement* also illustrates the productivity possible within these striated formations. Initially, the term hints at the etymological connection between assemblage and agency. It does so more than merely because of the common root, but because the French term indicates an active process of arrangement (not static). The "lines" of discourse and materiality are in the motion of arrangement. Within the territory, therefore, there is a constant play between and within expression (coding, semiotics) and action (pragmatics, passions). As my unicorn tattoo example illustrates, the mutations in the image on my shoulder demonstrate the tattoo as an active site of the relations between ink and skin, sun and air. The organized territory is not a static location, therefore, but an active site of the relations between expression and action, form and content. The "flow of agency" occurs within the territory as well as through lines of flight.[74] Moreover, *agencement* does not suggest that this arrangement is the purposive act of a human subject. Assemblage is not, as in the English definition, something that has been put into proper order or systematically arranged. It is, rather, the un-predetermined (though not chaotic or random) process of "fitting together."[75] The power to transform does not exist solely in the hands of the individual or collective actor who puts parts in their proper (or improper) places. Rather, transformation emerges in the processes of movement and intersection, mutation and (re)combination.

Assemblages of discourse (or, in Deleuze and Guattari's terms, enunciation)[76] function through processes of relative movement and stasis, flow and arrest. In these processes, discourses do not only act to assign meaning (though those moments do occur). Rather, they also have their own speed and intensity of movement that can intersect and collide. The movement in both axes operates at various speeds, with differing intensities, and through different dimensions. Because of this shift from object to process, the notion of assemblage fundamentally dismantles the representational economy of excess and exclusion:

> There is no longer a tripartite division between a field of reality (the world) and a field of representation (the book) and a field of subjectivity (the author). Rather, an assemblage establishes connections between certain multiplicities drawn from each of these orders, so that a book has no sequel nor the world as its object nor one or several authors as its subjects. In short, we think that one cannot write sufficiently in the name of an outside.[77]

It is not that books (images) and authors (artists) and reality (the world) do not have a discursive or material existence. Rather, it is that their objective existence and definitional difference should not presuppose or predetermine analysis. Instead, authors, art, and images are spectrums of diffraction (orders) enabling and constituted by various connections. Meanings are not, for instance, in *inter*textual or *inter*subjective relation to each other. Rather, meanings (or subjects and objects) are *of* relations; they are greater than the sum of their parts because they emerge out of the space of relation. These multiplicities function simultaneously as resources for other relations and connections. The shift in orientation, therefore, is to attend to the relation itself. Not the relation constituted by the variety of points of view produced by a priori subjects and objects intersecting and interacting, but the relation as the movement of, in our case, discourse, form, and content that occasionally sediments into always already multiple subjects and image objects. It is the apprehension of the truth of variation itself rather than simply recognition of multiple points of view.[78]

Agency in *agencement*, therefore, becomes the active power that exists in the merging together of actions and expressions, forms and contents, within the territory. There are possibilities for agency within the process of territorialization and performative reproduction as well, and not only in moments of deterritorialization or lines of flight. This approach emphasizes that territorialization and deterritorialization are not synonymous with liberation and oppression. Deleuze and Guattari are as clear as they want to be that territorialization is not inherently bad and deterritorialization is not always good. Both processes have their unique "dangers," be it homogenization or "pure and simple" destruction and death.[79] This approach also avoids the limitations for agency structured within a representational economy of resistance that I have already articulated.[80] Rather than reproduce another reading of tattooing as a form of social expression or as the inscription of power and discourse upon the human body, I want to look at what happens within these discursive operations. In other words, to look at the possibilities that emerge within the process of performative discursive reproduction and repetition.

One focus of analysis of the agency in territorialization is, consequently, on the glacial slowness that produces point-like moments of intersection (meanings) through the movement of reproduction and repetition. How, in other words, meanings or tracings get produced through the performative activity of discourse.[81] The agency of this machine (the potential power of transformation) occurs within the processes of repetition, reiteration, citation, and reproduction, within the various discursive and material formations and their strategies.

The other focus concentrates on the possibilities for agency within the visual image *as* a tracing. Not only, then, does the movement of territorialization function at the level of performative discourse, it also functions in the relations of form and content within the image. Deleuze and Guattari argue, for instance, that (some) visual images are tracings.[82] In other words, some forms of the visual image (i.e., an advertisement or a photograph) are, like the discourses used to interpret them, acts of territorialization. The space of the image is constituted by the various relations of form and content, visual and lexical registers. For instance, Deleuze's analysis of the Baroque emblem indicates that there are moments where the image exceeds or subverts meaning.[83] Tom Conley explains that "the relation of difference established between the visual and lexical register effectively exceeded the emblem's objective of control, subverting its moral or edifying aims. In Baroque emblems allegory is built upon a principle of excess. Images break free through their frames to 'form a continuous fresco,' hence to become *events* that multiply by virtue of a design."[84] The material and discursive constraints (form and content) of the image-as-trace actually produce the conditions through which the image acts out. Images can break free "through" their frames and not only despite of or in resistance to them. Any attempt to understand the activity within and of the image, therefore, requires attending to the image as a tracing.

There are, then, two levels through which the agency of the territorialized image works: at the level or plane of discourse, and at the level or plane of material form and content. These are not distinct operations, however, because the assemblage is a constantly moving network. Nor are "form" and "content" merely objects through which meaning is communicated and that can be consequently decoded via a semiotics of style. They are, in fact, lines of motion that can be slowed down or sped up.

ENCOUNTERING THE IMAGE

I realize that up to this point the description of agency via the language of assemblage and territorialization has been relatively abstract. This is primarily due to the fact that the agency of discursive destabilization and disruption is, by the nature of the assemblage, tied to the specific formal and discursive characteristics of the image and the meanings circulating through it. As Deleuze and Guattari suggest, there is a constant reciprocal play within the assemblage between expression (form) and action (reiteration and reproduction). Because agency is tied to the specific form and content of the image and to the particular discourses surrounding that form and content, recognizing those moments of destabilization requires an attention to method.[85]

Deleuze and Guattari's description of the movements of smooth and striated spaces within and through networks of assemblage calls for a radically contextual and historicized method of inquiry that does not reduce context and history to striated space. Their prose is necessarily abstract in its general description because anything less would begin to participate in the construction of categories that, when used as critical apparatuses, fail to account for specific, and I would argue, historical details. Deleuze and Guattari, for example, criticize linguistics and psychoanalysis for constructing "ready-made" categories that despise anomaly for adherence to the category.[86] Once you have decided, for instance, that everything has a meaning, then you will find nothing but meaning. If you decide, on the other hand, that you need to look for something that is an "other" to meaning, then you might look for emotion or materiality. In both cases, you are probably going to find either the presence or the absence of whatever category you are looking for depending on how you define and limit the category.

Context, in the unready-made sense of the assemblage, however, is not a locatable object that the critic subsequently describes. Similarly, history is also not a territorialized object only contained by sedimented time and space (past events, for example). Moreover, the relations between and among the assemblages of context and history are not necessarily linear (from past to present) nor cause-effect (the image is constructed in response to contextual factors or the past events directly lead to the present effects). Finally, history and context are not necessarily environments conditioning the behavior of subjects, but are assemblages within which complex relations produce subjectivities.

As a result of this particularly complex understanding of context and history, there are two primary goals for method. The first goal of the method is to make sure that history and context, form and content are not reduced to analytic categories but are viewed as processes. The second goal is to put the tracing back on the map. In order to address these goals and attend to the radical contextuality and historicity of the image as assemblage, I turn, as I suggest in the Introduction, to genealogies of performance. Performance genealogies approach history and context as contingent and reciprocal evolutionary processes of active and passive forgetting and remembering constituted in and through remnants of bodies and spaces. This idea combines a definition of performance as "restored behaviors" of concurrent, past, and future acts and events, with the practice of genealogical criticism adapted from Michel Foucault's work.[87] In order to follow the "legs" of images, therefore, performance genealogies provide a method of inquiry that perceives the style, character, speed, and direction of their movement. Genealogies of performance consequently attend to context and history, bodies and space as an assemblage and so offer a method for attending to the process of history and context as and through their smooth and striated forms. The motion of discourse, the motion within the image, and the motion of history are all, therefore, movements and performances of reproduction and reiteration within an assemblage.

This method also acknowledges the productive activity of the critic in creating the specific genealogy by means of critical interventions in images and texts. This approach to culture and cultural texts, therefore, offers an example of a critical method that combines dislocated interpretation with textual production in such a way that appreciates the complex and constantly changing cultural assemblage while attending to the specific characteristics of particular cultural and textual striated formations. In other words, this method allows the cultural critic to talk about a specific text (an image, for example) in relation to other past and present texts (images, literature, events) as well as to talk about how that text, its relations, and its lineage are changing, unstable, and are both a product and a producer of meanings, histories, and contexts. It is also a method that acknowledges (in fact, requires) that the act of criticism can go past diagnosis and description. Consequently, this method offers at least one example of a critical approach that escapes the "boredom" and predictability that is produced by criticism that repeatedly discovers that texts resist and reproduce dominant ideologies.[88] Critical genealogies of performance are one way to "construct media Texts from various fragments of our semiotic laden landscape" in order to offer a more productive and generative style of criticism.[89]

The second goal of method, as previously mentioned, is to put the tracing back on the map.[90] Images are territorialized moments of striated structure that emerge within the network of the assemblage. Consequently, describing them offers the first and necessary critical operation. The next move must be to perceive these tracings as entryways back into the movement of the assemblage. One way that this can be accomplished is for the genealogist to enter into the image in order to activate the historical and contextual assemblages and to perceive, through form and content, the movement of discourse within the tracing and, in so doing, to identify the potential for agency within territorialization. Deleuze and Guattari describe, for example, the haptic close-range vision of art where the "observer" does not stand back and look at the images but enters into them.[91] The form of entry I take in the analyses throughout this book is to attend to the movements of the bodies within the images. Not only in terms of their actual movements (as in performance), but also in terms of the movements remembered. I enter into the assemblage or map of history, for instance, through the gateway of the image. I enter into the assemblage of space and place through the gateway of form and content.

Putting the tracing back on the map also means being responsible to the ethical and political realities of the processes of reproduction and repetition. The bulk of the analyses in this book are primarily concerned with the tracings (images, meanings, discourses) themselves and the potential emergence of disruption within territorializing actions (interpretations, judgments, evaluations). That is, considerable attention is paid to identifying and interpreting the various classed, gendered, and other such meanings associated with tattooed bodies and to how those meanings are discursively produced and structured in order to carefully describe the specific terrains through which the images might potentially act disruptively. To take the potential of image vitality seriously also means being responsible, however, for the normalizing and oppressive effects of capture that these interpretations and meaning attributions can perpetuate: "That is why it is so important to try the other, reverse but nonsymmetrical, operation. Plug the tracings back into the map, connect the roots or tree back up with a rhizome."[92] I interpret Deleuze and Guattari's suggestion as a request to use the analytic process to generate other "as-if" potentials instead of stopping at the point of deconstruction and decomposition. Consequently, each chapter's description and interpretation of meanings, discourse, and discursive strategy is followed by an effort to pursue the potential lines of flight or generative possibilities produced through the critical encounter with the image.

As I hope has become clear through the discussion in this chapter, this is a book about the assemblages of visual imagery, history, signification, and performance and the possibilities for agency within territorialization. The following chapters address those themes by exploring the restored performances of performative reproduction through very precise analyses of the form and content of specific images as territorialized gateways into the assemblages of context and history. Throughout, the tattooed body is offered up as a specific example of the ways in which meaning operates, how it can be transformed, and, most importantly, the possibilities for transformations that are "other" than, but dependent upon, meaning. The form of analysis follows Roach's suggestions by describing the images and image events order to provide a picture of the "present" performances. I then proceed to map the historical events and discourses that the image's form and content revises, recites, and relies upon for its social and cultural meaning. The resulting documentation therefore simultaneously shows the influence of history on interpretations of contemporary tattooed bodies on display as well as illustrates the ability of the contemporary context and performances to call historical authority and the construction of cultural memory into question. In so doing, the analyses demonstrate the productive potential of contemporary performances and representations to actively construct and transmit (new) cultural memories through the agency of transformation within the performance of performativity.

CHAPTER TWO

MODERN DAY SIDESHOWS: CLASS, CLASSIFICATION, AND TATTOO CONVENTIONS[1]

I HAD JUST TURNED 21 AND GRADUATED FROM COLLEGE. I was about to begin the long road trip south to begin my graduate student career and was celebrating with friends over some good-old-fashioned Bud Light and live classic rock covered by a local band in what might generously be described as a "dive" bar. During the course of the evening, one of my close friends began to talk about how she had always wanted to get a tattoo. "Me, too!" I chimed in. As luck would have it (as is usually the case in narratives such as this), a nice gentleman with long blond hair wearing a black leather Harley vest exposing extensively tattooed arms overheard our conversation and offered to make an appointment for us the following morning with his tattoo artist. Ten minutes and a pay phone call later, we had our appointments. At 9 a.m. the next (same?) day, I acquired my first tattoo. Becoming tattooed heightened my awareness of my body as a site/sight of identity and class evaluation. Since then, I have become increasingly curious about how the tattooed body both metaphorically and literally illustrates the complexity of the sometimes simplified notions about the social inscription of class norms upon bodies, and of the location of classed subjectivity in the external expressions of an interior self.

In this chapter, I analyze the visual display of tattooed bodies in terms of how class meanings accumulate over time and of how these visual displays resist and recuperate, rearticulate and reproduce meaning. Moreover, I am interested in the accretion of class meaning across and through historical contexts, situating the performativity of the image of the tattooed body's performance within and through its citation of history. The purpose of this chapter is to address the issues of class inscription and materiality, and the connection between visual expression and subjectivity, through an analysis of visual displays of explicitly tattooed bodies in tattoo contests held at tattoo conventions through a critical genealogy of performance.

Through this analysis I suggest that contemporary displays of the explicitly tattooed body in the contest context rely upon and reiterate historical practices and discursive strategies mobilized to contain and classify the excess of the explicitly tattooed body. Within the highly saturated and heightened visual physical and spatial context of the tattoo convention, the explicitly tattooed body on display inverts and seeks to appropriate classification discourses and practices that have historically acted to cast it as deviant and low class. This tactic is, of course, both successful and not successful in resisting the effect and force of dominant ideologies. Rather than dwelling on the simultaneously resistant and recuperative effects of the contest performances of the explicitly tattooed bodies, however, the analyses focus on *how* the territorialized visual image of the body in these performances cites historical discourses and how that citation is necessary for both resistant and recuperative effects.

In the end, the force of normalizing class discourses and historical citation to contain and dictate the (deviant) meaning of the explicitly tattooed body is challenged through actual and immanent *movement.* The discourse of classification does not fully contain the agency of the explicitly tattooed body's image because of the capacity of the image to move performatively *within* this discursive territory. I conclude, then, by emphasizing the potential for agency, which is always immanent in the image of the body, by showing how the explicitly tattooed body's actions at the contest dismantle the logic of normalization even as they resist and recuperate it. It is, in short, in *movement* (action, production, performance) where the greatest potential for the disruption and (re)articulation of, in this case, the deviant/normal discursive dialectic exists.

While it might appear unusual to begin a study of visual imagery with an analysis of a performance context, this chapter, in addition to exposing the class evaluations of the tattooed image, establishes the relevance of the human body to this study of visual agency. As I indicated in the Introduction, the choice of the tattooed body is significant for its metaphorical force as a body. In this chapter, the contours of that force are made explicit through the analysis of the activities of the tattooed body in "live" display. Additionally, this chapter takes snapshots (traces) of the tattooed body within the performance context. The analysis of the performance is an examination of the visual image (as defined in the Introduction) of the tattooed body *within* the performance. This chapter therefore makes explicit the territorializing process of discursive framing as well as lays the groundwork for the power of the image to disrupt the territorialized frame.

CLASS PERFORMATIVITY

Throughout this analysis, I approach class as a performative construct. Tattooed bodies, moreover, hyperbolize discourses of class and status. The narrative introduction to this chapter, for instance, calls upon several of the stereotypical signs of (lower-) class performances associated with some tattooed cultures and bodies: cheap beer, classic rock music, dive bar, Harley-Davidson vest, and the like. Class is produced by the performance of the body in cultural space and practices of the body are reciprocally interpreted as measures of social status. Bourdieu articulates the reciprocal relationship between production and interpretation in terms of taste: "In the ordinary situations of bourgeois life, banalities about art, literature or cinema are inseparable from the steady tone, the slow, casual diction, the distant or self-assured smile, the measured gesture, the well-tailored suit and the bourgeois salon of the person who pronounces them."[2] Embodied visual displays are signs of social status, so it is not enough to utter the correct pithy statement unless the entire picture and performance is intact. Moreover, because the social elite sets the standards for tasteful and appropriate behavior, they guarantee their position at the top of the social hierarchy. Performances and images that do not conform to norms established by the upper class identify the social status of the performer as both deviant and low class. Social performances that challenge the assumptions of upper-class social norms are, consequently, devalued as "vulgar and tasteless."[3]

Class evaluation of the body extends beyond taste as an external symbol into the realm of the internal. Society interprets norms inscribed upon the body, either literally or figuratively, as symptoms and clues of individual subjectivity.[4] Bodies violating social norms, therefore, are interpreted as external manifestations of internal social deficiencies. Evaluations of working-class behaviors and images, for example, produce stereotypical labels such as "tacky, unhealthy, hopelessly bigoted, and parochial."[5] More than simply demonstrating economic status, therefore, class norms signify one's physical condition, psychological well-being, and intelligence. Even if the working-class individual achieves the material means required to rise above his or her economic class, social status simply shifts from "tacky and trashy" to "tacky and trashy with money." In their discussion of "white trash," Newitz and Wray indicate, for example, that this label "is a way of naming actually existing white people who occupy the economic and social margins of American life, and it is a set of myths and stereotypes that justify continued marginalization."[6] Economic class may be a material reality, but status is communicated by stereotypical evaluations of visual imagery that measure individual worth and social value.

Styles of physical adornment figure prominently in class evaluation by marking the body's social status and making it culturally visible.[7] Clothing, makeup, hair styles, and accessories all communicate some version of class identity, as any casual glance at popular film or television programming will demonstrate. Tattoos are a unique version of adornment because they appear to be a relatively permanent physical inscription. This permanence both resists the possibility to shop for class identity through changes in adornment as well as confronts the belief that the self is not of the body but of the spirit or mind. In her explanation of Alphonso Lingus's discussion of tattooing, Elizabeth Grosz explains,

> It offends Western sensibility (at least the white, and especially middle-class sensibility, although Lingus doesn't specify this) that a subject would voluntarily undertake the permanent inscription of a verbal or visual message on its skin. Its superficiality offends us; its permanence alarms us. We are not so much surfaces as profound depths, subjects of hidden interiority, and the exhibition of subjectivity on the body's surface is, at least from a certain class and cultural perspective, "puerile" (his [Lingus'] word).[8]

Tattoos, from this point of view, represent an immature and shortsighted decision. They irresponsibly mark a stage that a person might be going through that s/he will eventually grow out of and, presumably, regret. Unlike temporary forms of adornment, which emphasize the idea that one can both reveal and conceal one's "true" self, tattoos cannot be discarded at the whim of the wearer, in response to a shift in social opinion or, as the following discussion explores, because of a change in geographic location.[9] Unlike, for instance, class slumming (popular in the late nineteenth and early twentieth century in the United States and in Britain where upper and middle class men and women would visit slums in search of entertainment and then return safely to their homes once their curiosity was satisfied), tattoos cannot easily be abandoned. The conceptual dilemma tattoos evoke through their explicit and "unnatural" visibility, therefore, is between the modifiable subject that can shift identity or perform different roles, and the fixed and relatively stable subject having chosen to express its deviance.

Tattooed bodies ultimately, therefore, force a visual recognition of the classed performativity of the body. Unlike other forms of social inscription that can be hidden because they have become naturalized (the codes have become invisible), or altered because they are temporary, tattoos constantly confront naturalization because they signify a permanent decision to mar the pure surface of the body. Metaphorically and literally, tattoos illustrate the inscription of social norms and codes upon the body. Class evaluations uniquely impact the tattooed body, therefore, because it remains tattooed regardless of where it is or how it is culturally evaluated. Tattoos implicate the skin; they become part of the body assemblage, and therein lay their power and the subsequent desire to territorialize their meaning.

Class evaluations of the visual image of the body also depend upon the social and cultural space of display. There are places where images of the tattooed body are more or less accepted. Cultural space is socially coded and constructed to encourage the presence of some bodies and images and to reject the intrusion of others. Sibley suggests, for example, that the structure of space and spatial relations in any given society will reflect the strength of class boundaries. Highly stratified (or striated and closed) social systems will produce spaces that have clearly delineated (strongly classified) boundaries welcoming some types of behavior and refusing others.[10] Social systems that are not as strictly stratified (more open or smooth), however, will produce weakly classified spaces that allow intermingling of social classes.[11] The image of the tattooed body, as we shall soon see, has the capacity to act out of strongly classified space as well as to effect mutation and so alter weakly classified space.

In strongly classified/striated spaces, the image can potentially disrupt spatial structure by inventing new possibilities for interaction within the territory. At an art gallery that I recently visited, for instance, there was a strongly classified display of twenty framed artworks—all of the same square shape and size (approximately 1' × 1') and displaying a similar palette of muted brown, black, and copper tones. These pieces were hung in two parallel and even rows. The spatial structure suggested that the gallery audience ought to move through the display from left to right, choosing (or not) to purchase an item once they arrived at the right end where the catalogue of prices was displayed on a pedestal three or four feet tall. Despite this organization, however, the form of the works themselves invited a different response. These were mixed media studies of human bodies that had sculpted miniature human heads attached to material objects such as cork screws and pliers. The faces seemed to jump out at the viewers, inviting them into the display in ways that defied the classified space of their organization.

Weakly classified smooth space, on the other hand, invites the presence of multiple classes. A space of intermingling such as the subway in New York or a train station, for instance, suggests relatively weak classification. Weakly classified form and content in visual display could be achieved through practices of found art assemblage, for instance, where artists combine natural and "made" found objects into pieces of artistic expression. Weakly classified space consequently is open to intermingling though not *necessarily* to mutation. The forms that interact do not necessarily change through their interaction because the lines of flight may occasionally merely intersect. In the subway, for instance, the construction worker's brush against the New York University student does not contaminate either class position through this contact. As another example, in Joseph Cornell's work that assembles objects in three-dimensional boxes, the items remain distinct in their combination.

In order to articulate the importance of the tattooed body and its performative and historical context to class and status interpretations, the following discussion enters from two different directions into the image of the tattooed body in the tattoo contest at the convention. The first encounter evokes how the bodies in the tattoo competition kinesthetically recall the style of presentation common in the late nineteenth- and early twentieth-century sideshow performances of illustrated people that took place in popular amusement venues such as dime museums, carnivals, circuses, and world's fairs. As a result, the tattoo contest bodies and context subsequently also appropriate the scientific logic of physical classification that circulated around sideshow performers in the late nineteenth century. Rather than treat tattoos as signs of low-class criminality and deviance, physical classification in the contemporary context exalts the aesthetics of tattoo form and the skill of the artist. Contemporary display on the competition stage thus inventively displaces the authority of scientific classification by manipulating its interpretive norm; classification strategies turn the tattoo into something to celebrate as opposed to denigrate. Within the tattoo community, however, the valorization of tattooing's artistic form precipitates internal class stratification.

The second encounter is with the relations of space. I argue that the tattoo convention's spatial configuration actively remembers the social organization of space at sideshow venues and their position in/on the threshold of society. Both historic and contemporary performances are situated within a highly classified and territorialized organization of social space that facilitates the inversion of status hierarchies through the destabilizing force of the weakly classified threshold. In this case, smooth spaces emerge within and, more importantly, because of highly striated territorialization. The tattoo convention behavioral vortex produces a context, therefore, where the tattooed and deviant body is celebrated and the nontattooed body is out of place. This inversion is, however, contingent upon contextually specific spatial and temporal conditions.

Tattoo Competitions as Modern-Day Sideshow

Since the 1970s and the development of professional tattoo organizations such as the National Tattoo Association, tattoo conventions have become a worldwide phenomenon where tattoos, tattooing, and tattooed bodies are celebrated.[12] Tattoo enthusiasts attend conventions held in places from

New York to Los Angeles and from Tokyo to Amsterdam. Tattoo artists and collectors gather at these conventions to conduct and participate in workshops, to advertise and sell tattooing supplies and equipment, and to tattoo and be tattooed. Tattoo conventions constitute a space for the enactment of tattoo community.[13] Tattoo artist and enthusiast Madame Chinchilla views the convention as "an occasion to mix with cohorts, see and show tattoos, enter tattoo contests, go to seminars and collect a tattoo as a memento of the event. It is also an opportunity to be photographed, interviewed, and to communicate on like-minded subjects."[14] Tattoo conventions provide the tattoo artist and tattoo consumer with the opportunity to gather together and celebrate their bodies and art.[15] In fact, conventions have become the "central medium of cross-pollination. Artists can see what other artists are creating, either in person or through the evidence of the living flesh."[16]

A premier event at each of these conventions is a "live" tattoo competition, pitting the talents of the various artists, and the tattooed bodies in competition, against each other.[17] Tattoo competitions are venues where tattooed bodies parade across stages and have their tattoos and bodies evaluated by a judging panel normally composed of prominent tattoo artists and personalities. The contestants and the tattoos they display are evaluated based upon "their tattoos, the appearance of their bodies, and the status of the tattooist."[18] The tattoo contest is a venue within the convention that heightens the behavior of the tattooed body by placing the body on stage and before an audience for the express purpose of being evaluated as a tattooed body.[19]

Between 1999 and 2004, I attended four tattoo conventions. On Halloween weekend in 1999 and 2000, I attended the Southern Comfort Tattoo Expo (also known as Tattoo Voodoo) held at the Landmark Hotel in Metairie, Louisiana. In May of 2004, I traveled to New York for the 7th Annual New York City Tattoo Convention held at the Roseland Ballroom on West 52nd Street (figure 2.1). In June of the same year, I found myself at Live Fast: The Houston Hot Rod Tattoo and Kustom Kulture Art Show. More than the mere pleasure at the chance to acquire another tattoo, however, attending these conventions was an opportunity to explore the representation and performance of the explicitly modified tattooed body.

Illustrated Performances

Tattoo Voodoo takes place annually in the Mardi Gras ballroom located on the lobby level of the Landmark Hotel.[20] The ballroom is a long, red-carpeted, glass-chandeliered rectangular room. Booths were set up along three walls and down the center of the ballroom. Each booth hosted one or two artists/studios. Inside the booths, the artists were talking to each other

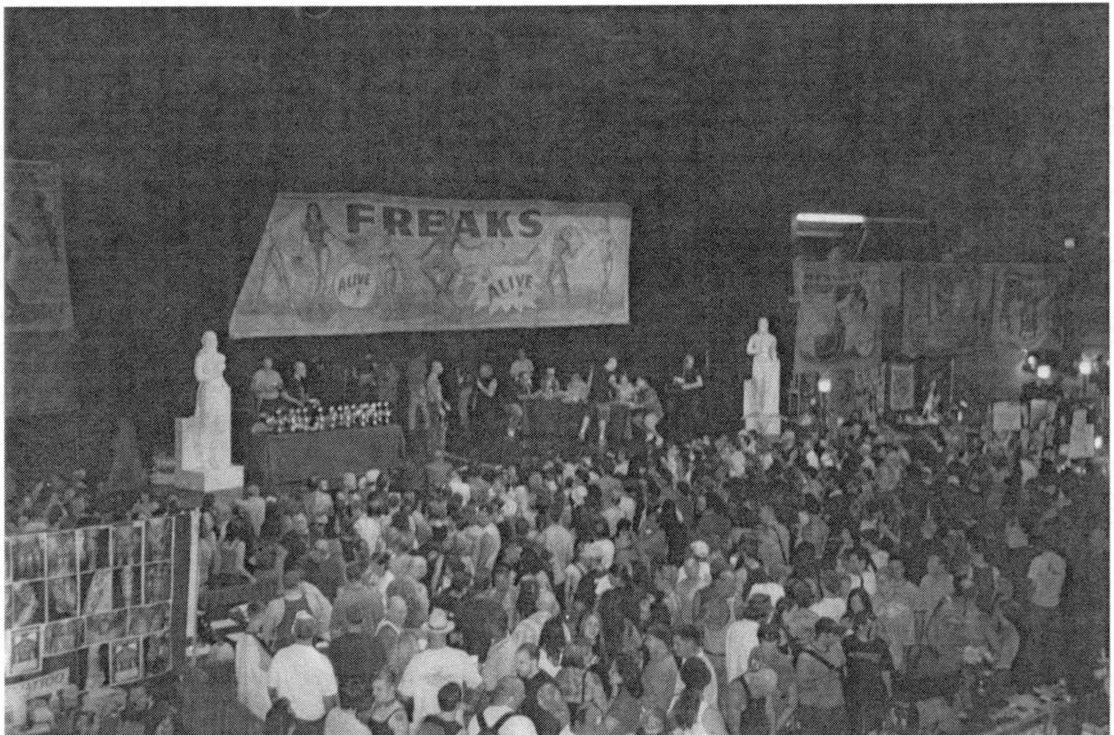

Figure 2.1 2004 New York City Tattoo Convention.

Source: Photograph Mindy Fenske.

and to the potential customers and other artists meandering by, drawing and performing tattooing. The back wall of each booth was usually decorated with a large poster or cloth mural of artwork with the name of the studio and/or its artist(s). Adding definition to the aura of the space was the background hum of tattoo machines, classic rock music, voices in conversation, and an occasional announcement over the sound system (usually reminding attendees and participants of contest entry deadlines).

The tattoo competition at Tattoo Voodoo in both 1999 and 2000 took place on a stage at the rear of the Mardi Gras ballroom. At the back of the stage stood a table displaying the trophies that would be presented to the victorious. The three extensively tattooed judges of the event sat at a table to the right of the stage. The tattoo competition entrants waited outside the building and entered when their turn came through a door immediately to the right of the judges' table. An M.C. with microphone in one hand and a clipboard in the other would announce the category, the name of the artist, the name and location of the artist's studio, and the name of the person whose body was the canvas for the artwork. When his or her name was called, the tattooed contest participant would proceed to the table of judges. At the judging table, the entrant presented the tattoo to the judges for inspection. The judges would scan the tattoo, jot some notes down on paper, and then nod to the M.C. to call the next entrant. The performance of the tattooed person in the tattoo contest was both in homage to the artist's prior performance, and in honor of the tattoo collector. Artistic skill and aesthetic form of the tattoo art were evaluated and the prize awarded to the tattoo collector and the tattoo artist. The tattoo competition combined the aesthetics of a fashion show with that of an art gallery. Competitors became models because their skin was the artist's canvas.

The men in the competition generally wore t-shirts or tank tops or were shirtless with jeans or cut-off jeans shorts. They ranged in age from twenty-something to sixty-something. The women displayed more variety of attire, wearing anything from black leather pants and halter-tops to cut-off jeans and t-shirts. What was not common for the men was any version of pressed khaki and tucked-in, button-down, cotton shirts. The women's styles, although not so easily described in detail, tended to be either jeans and t-shirt casual or leopard-skin fantastic. In all cases, the clothing worn on stage conspicuously revealed the tattoo in competition. For example, the woman on stage wearing shiny black leather pants and a red, tied around the neck, spaghetti-strap halter-top was featuring a large dragonfly tattooed on her back that stretched below her waistline. The full tattoo was visible because there were no straps on her back and her pants rested low on her hips.[21]

The norm was for the individual to walk to center stage and stand passively in a pose that allowed the audience to see the tattoo entered in the competition. The stage frame territorialized the active and moving body into the canvas for tattoo display; it was the tattoo, and not the tattooed body, that was the focal point of this performance.[22] For example, a man with a tattoo on his throat threw his head back, another gentleman to display his tattooed calf placed his leg in front of him and bent his knee so that his leg pivoted on pointed toe in a manner reminiscent of shoe mannequins in department store displays. A woman who was revealing a tattoo in her pubic region unzipped her pants just far enough to display the artwork. The performance was a display of artistically classified body parts. The parade of legs, arms, chests, backs, necks, pubic areas, and even penises continued until all entrants had crossed the stage. This tattoo classification system territorialized the body by enacting a Frankenstein-like dissection of the human body for tattoo inspection.

The conventions of display within this event in particular and at the convention in general also established physical exposure and exhibition as the norm. Madame Chinchilla notes, "The attire is provocative; ovals and designs are cut out of clothing to show-off tattoos."[23] Or, as Krakow describes, "there's usually a fair amount of tasteful nudity."[24] Throughout the convention, it was also not unusual to see people pulling and tugging at clothing items to reveal tattoos on stomachs, backs, and upper thighs. Clothing that partially revealed and partially concealed tattoos was shoved aside so the viewer could get a full view of the design. Not all convention attendees were wearing revealing clothing, of course, but the norms of physical display certainly encouraged exhibition. As Braunberger's analysis of tattoo conventions suggests, the "exhibitionist melancholy to share one's tattoo" is eased and "strangers touch each other as they admire tattoos, they tug clothing away; most wear little to begin with."[25] Tattoo conventions are safe havens where tattoo artists and tattoo collectors gather together to compare tattooing techniques and to show off their tattoos.

In order to understand how such an event and environment might be possible, a historical detour is necessary. This parade of tattooed bodies featuring tattooed body parts on the stage at Tattoo Voodoo, as well as the trend toward physical exhibition among those not entered into competition, physically rehearses the style of presentation of illustrated men and women at the turn of the twentieth century when the popularity of exhibitions of the tattooed body was reaching its peak in the United States. Illustrated (extensively tattooed) men and women occupied freak show stages in venues from dime museums, which displayed a wide assortment of human anomalies and oddities, to the sideshow tents at stationary and traveling carnivals and circuses. Audiences to these museum displays could find tattooed bodies exhibited next to Siamese twins, bear ladies, and dog boys. The popularity of these tattooed exhibits in the United States lasted from about 1840 until the 1930s.[26]

More recently, there has been a resurrection of sideshow performances through venues such as the Jim Rose Circus, and individual traveling acts such as the Human Marvels featuring Enigma and Katzen.[27]

The first kinesthetic recollection of sideshow performers is the tattoo competition's evaluation of the tattoo as a sign of the skill of the tattoo artist. The tattoo, within this context, is the record of the artist's performance of his/her craft. Similarly, the performance of the tattooed freak was, in many ways, already accomplished before the actual exhibition of the body as a product was put on display. As Hammer and Basker indicate, the "appeal of the tattooed freak was the ability to withstand pain and bizarre acts of mutilation. The tattooed performer appeared as one who had defied pain, not in one great gesture (like a nail in the nose) but in a thousand acts of bodily violation."[28] In the early sideshow tents, it was not necessary for illustrated people to do much of anything because their bodies were the physical evidence of the event already accomplished and endured; they were living picture galleries.

The event that the illustrated body served as verification of, moreover, was often a fictional narrative of captivity. For instance, the very first recorded tattoo exhibit in the United States, James O'Connell in 1840, sold a pamphlet detailing his worldly travels and experience of being forced to undergo tattooing at the "hand of indigenous maidens."[29] The actual physical performance of the tattooed person achieved meaning because the images etched upon their body provided evidence verifying the "truth" of these often fictional narratives surrounding the acquisition of the tattoos. As a result, the tattooed performance in early display generally consisted of the seated or standing tattooed body juxtaposed with the narrative explaining the circumstances of becoming tattooed.

According to Richard Bogdan, for example, "Captain Costentenus" was P.T. Barnum's first illustrated man, and his performance is generally characteristic of early displays of illustrated people. Costentenus' hair was curled and braided and fastened to the top of his head. He wore a loincloth and a diamond ring. During the performance, he chain-smoked cigarettes.[30] Fellow circus performer Robert Sherwood's account additionally suggests that the Captain began his performance fully attired before stripping down to the loincloth: "the temperature under a wet circus top in the early spring is sometimes of such a nature as to make a well-clothed man shiver; how much worse, then, if one had to strip a dozen times a day to seminakedness."[31] Unlike other sideshow exhibits—who either performed tasks one might assume could not be done by a person with their particular disability (armless individuals painting, for example), or who engaged in more standard show business (singing, dancing, piano playing)—the illustrated person merely had to sit or stand nearly nude upon the stage.[32] It was not, consequently, the animated body that was of interest. Instead, the illustrated exhibit was a combination of painted body parts on display.

Performances of illustrated women followed similar conventions of display. The fundamental difference was that these women were not only violating general norms for proper physical exposure, they were trouncing upon Victorian limits of what was deemed acceptable for women. The following description (in the *New York Times*) of Irene Woodward's March 18, 1882, appearance at The Sinclair House demonstrates both the style of performance and the norms violated:

> During a reception of three hours at The Sinclair House yesterday afternoon she was attired in a scant costume of black velvet and gold. A close-fitting bodice or jacket, trimmed with gold bullion and fringe, surmounted a pair of trunks whose long golden fringe stopped an inch or two above the knee. . . The visitors were permitted to look upon the quaintly decorated skin of the upper portions of the chest and back, the arms and the exposed portions of the lower limbs. Miss Woodward remarked that she felt a little bashful about being looked at that way, never having worn the costume in the presence of men before.[33]

Woodward (figure 2.2) was merely required to sit and be gazed upon because she was wearing so little. Illustrated people, at least until the 1920s, simply showed their skin.[34]

This turn-of-the-twentieth-century sideshow mode of presentation was clearly reenacted by many of the entrants in the tattoo competition at the Tattoo Expo during the turn of the twenty-first century. The men and women on the contest stage, like their early century cohorts, were there to model their tattoos, not because they were expected to *do* anything at all. In both the context of the sideshow and the tattoo competition, the tattoo is evidence of a prior performance. In the sideshow, the prior performance was the event of becoming tattooed, which was often described as the now illustrated person being forcibly tattooed after being captured by some vague "native" culture.[35] In the tattoo competition, however, the prior performance is that of the artist who designed and rendered the tattoo. In the context of the tattoo contest, therefore, the tattoo is recoded as a mark of aesthetic status, not of freakishness. In the late twentieth-century context, the entrant into the tattoo competition is living evidence of the skill of the tattoo artist and the aesthetics of the body. In a very compelling sense, the late twentieth-century body at Tattoo Voodoo is a human picture gallery transformed into an art gallery representing an effort to recode deviance as distinction.

The individual performances of a few of the Expo tattoo competitors, however, violated the behavioral norm authorized by their early twentieth-century

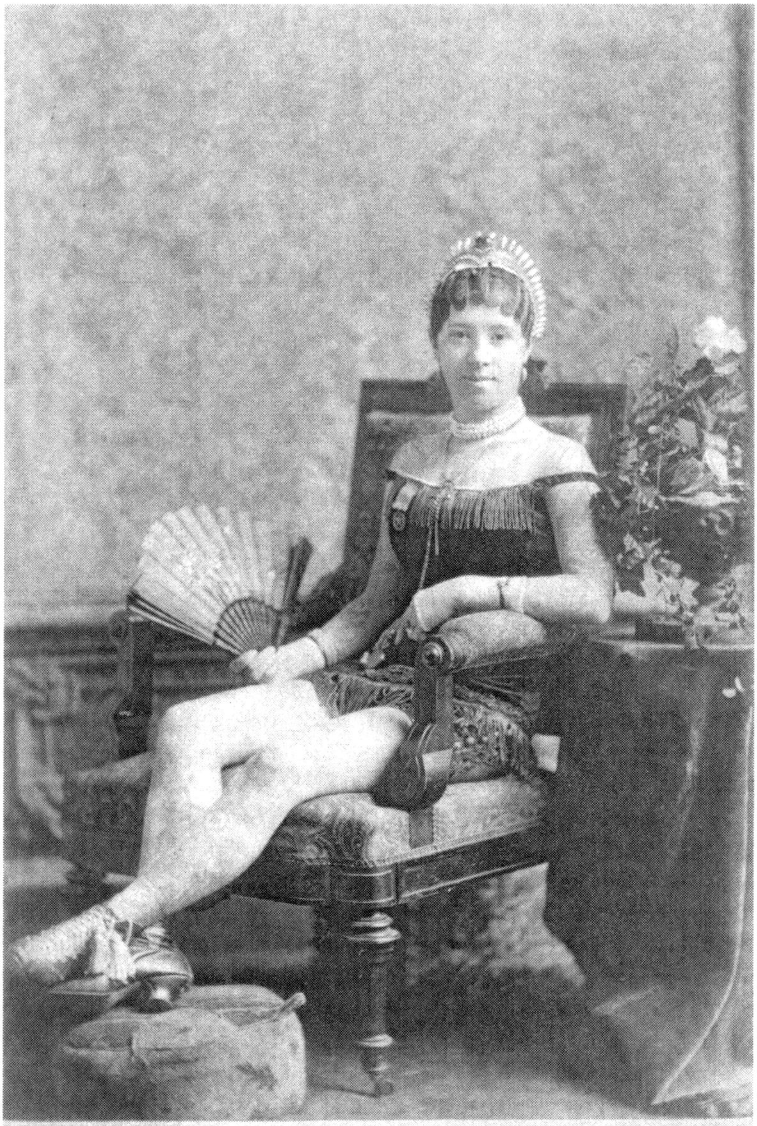

Figure 2.2 Irene Woodward. Tattooed Lady.

Source: Ronald G. Becker Collection of Charles Eisenmann Photographs. Special Collections Research Center, Syracuse University Library.

illustrated predecessors. For example, a slender auburn-haired woman wearing a scant bikini top and cut-off jean shorts stood center stage to display the tattoos covering her upper chest, arms and back. She stood passively, without expression, facing the crowd with her arms flexed in a pose featuring her biceps. She then slowly rotated her torso so that her back was to the audience before facing us again with a small, almost bashful, smile. At this point the largely male crowd went wild with catcalls. I think the catcalls were only partially the result of the woman's body on display. If it were solely because of her seminudity, all of the women would have received the same attention, but they did not. It was the incompatibility, the inconsistency between the diminutive smile and the muscular bodybuilder pose that resonated with the audience and produced (the possibility of) their response. The inconsistencies both within and between these almost simultaneous movements broke through the territorialized norm of passive display and, because of the particular way in which this was accomplished, subtly altered the context of the contest and so activated the potential for alternative movements—something like the inspiration for improvisation in jazz performances coming from the slightest change in rhythm. The man who followed this performance, riffing off the new vibe, shook his tattoo decorated pasty white belly. His belly's jelly-like performance was rewarded with applause and laughter.

Prior to these events, the tattooed body on stage conformed to the highly classified performance norm established by its early-century counterparts. The focal point of the performance was the art etched upon the body part. The skin of the stomach or the back was merely a territorialized canvas for the tattoo. The woman's particular movement, however, resonated in such a way that it produced a new effect in the audience because it acted out of the classified space of the gallery display. The audience's catcall response then dopplerred together with her movement, thus producing through this combination more possibilities for responses and effects.[36] The man's performance (made possible through the accumulated effects of her movement and the audience's response) then joined in, again altering the scene and participating in the production of new possibilities. In his case, movement transformed the canvas into a jolly belly and the audience laughed at the carnivalesque display. Much like photographs of tattoos that center the tattoo in the frame and sever the remainder of the body (somewhat like how surgical blankets depersonalize the patient by framing the incision point and masking the rest of the body), the context of the tattoo competition framed the body part and attempted to erase the person. But, as Judith Butler argues,

> [B]odies never quite comply with the norms by which their materialization is impelled. Indeed, it is the instabilities, the possibilities for rematerialization, opened up by this process that mark one domain in which the force of the

> regulatory law can be turned against itself to spawn rearticulations that call into question the hegemonic force of that very regulatory law.[37]

When the tattooed body on stage moved within the framed constraints of the competition display, it rematerialized. The possibility for movement was produced because of the strength of the striated frame to classify the body. Normative discipline became the platform upon which the body could act otherwise. The effect of this disruption also reverberated through the audience, compelling the awareness that they were staring at people as works of art. Moreover, the undulating wiggle of the pasty belly and the laughter it generated self-reflexively poked fun at the serious tone set by the organizers of the event (another ripple, another vibration). The result shook the frame of body as art gallery, produced laughter and catcalls, and deftly illustrates the capacity of the contextual frame of the tattoo competition to enable the resistant potential of movement always present in the living human body.

CLASSIFICATION AND CATEGORIZATION

Not only does the (not always) passive style of tattoo contest performances rehearse their early twentieth-century counterparts, but tattoo competition categories also remember the cultural performance of the freak show and, in so doing, cite scientific classification discourses dependent upon linking the exterior of the body to hidden interiority. As the following discussion reveals, the illustrated sideshow performer was subject to scientific, as opposed to artistic, scrutiny.

The tattoo competition separates tattoo styles into categories delineated by artistic style and body part. Competition categories highlight form, content, and location of the tattoo.[38] The following list of categories is an excerpt from the Tattoo Voodoo 2000 program:

> LIVE TATTOO COMPETITION
> The live tattoo competition will be judged by a panel of highly qualified tattoo artists. All contest winners will receive a first place trophy. The artist will receive a certificate of merit.
> There is an entrance fee for the live tattoo competition. This is to defray costs and to keep it serious.
>
> CONTEST CATEGORIES
>
> | | |
> |---|---|
> | Best color tattoo | Best black grey |
> | Most realistic/portrait | Best Full Back |
> | Best front panel | Best Native American |

Most unique theme	Best tribal
Best overall male	Best overall female
Best sleeve	Best single piece
Best tattoo of the show	Best booth

In the tattoo competition these categories of body parts and aesthetic styles are meant to provide a mechanism to evaluate the art of the tattoo. The body is visually fragmented into classifiable parts in order to be artistically evaluated. This system of classification and fragmentation of the body kinesthetically recalls and discursively reproduces procedures of tattooed body fragmentation and evaluation beginning in the late nineteenth century.

With the general advent of the social sciences, the arrival of criminal anthropology, and the building desire to classify and categorize abnormalities, turn-of-the-century freaks were increasingly under medical and scientific investigation as examples of deviance and abnormalcy rather than as fantastic specimens of wonder and awe.[39] For example, upon witnessing the body of Captain Costentenus, Dr. Kaposi, a professor at the University of Vienna, suggested in 1872 that "it would be desirable to form a committee consisting of specialists" in the disciplines of "medicine, ethnography, linguistics, art, and perhaps also archaeology" together with "anatomists, zoologists, dermatologists and surgeons who would be competent to examine and make a scientific study of the skin of the tattooed man Captain Costentenus."[40] Bodies in competition venues, be they beauty pageants or the Mr. Olympia contest, suffer the scrutiny of expert panels deemed qualified to determine their worth. The authority of the tattoo contest's panel of experts and tattooed personalities depends upon this tradition.

A defining characteristic of the evaluation process is the production of body part categories. In fact, Kaposi's own evaluation of Costentenus reads much like a catalogue (or the Tattoo Voodoo program):

> On the rest of his body I counted 386 of the larger figures, which, if one included the figures on his forehead, make a total of 388. These are distributed on the various regions of his body as follows:

chest	50
left arm	51
right arm	50
back	37
neck	8
hips and pelvis	52
penis	1
legs	137
forehead	2
total	388[41]

The scientific community in the late nineteenth and early twentieth century discursively dissected the tattooed body. Importantly, the tattooed parts were coded as forensic evidence of delinquency. Tattoos were signs of "puerile" interiority. The influence of Cesare Lombroso was integral to constructing the association between tattooing and criminality.[42] Lombroso "engaged in cataloguing physical characteristics which were supposed to signify mental disability, mental illness, immorality and criminality."[43] According to Bogdan, Lombroso argued that tattoos "were the stigmata of a distinct different form of human, the 'criminal man'"[44] The body part catalogue was, consequently, the criminal anthropologist's effort to represent the body parts of a tattooed person as documentation of the social depravity of tattooed people. The tattoo, consequently, came to be a sign of deviance. Several studies were subsequently conducted to test this theory.[45] Some of these projects merely sought to quantify the popularity of tattooing within specific communities (particularly the Navy), while others desired to verify the claimed connection between tattooing and social deviance and delinquency. The tattooed body itself was coded as an enactment of primitive, lower-class, and criminal behaviors. The image of the tattooed body was, to borrow from Bakhtin, consummated and finalized as a lower class form by the gaze of the scientific community.[46] It had become a trace.

The act of tattooing visually confronted and threatened to disturb notions of the controlled and natural purity of the surface of the body. Scientific discourse responded to this potential deterritorializing threat with strategies of control, thus transforming the potentially smooth space of the body into a deviant striated formation. Mary Douglas explains this territorializing process as an attempt to make order out of an untidy situation.[47] The body that does not conform to norms of physical homogeneity and behavioral control is constructed as deviant in order for society to maintain a sense of order and to assert control. In *Purity and Danger*, Douglas articulates these issues in terms of purity and pollution. She argues that purity is associated with that which is clean, controlled, unmarked, and healthy. Polluted and, consequently, devalued and lower-classed bodies are unruly and marked. Within a hierarchical system of social classification, therefore, the image of the pure, clean, and controlled body that is "naturally" unmarked or unmarred occupies the preferred position. Conversely, the physical body that is disorderly and physically marred or marked is socially constructed as devalued and deviant. Social class is assigned to a body, therefore, by how its visual performance enacts social norms of purity or pollution. Within the context of late nineteenth-century U.S. culture, physical exposure and conspicuous self-display accordingly became signs of lower-class delinquency.[48]

The legacy of scientific evaluations continued into mid-twentieth century. Sanders indicates that by this time "tattooing was firmly established as a definedly deviant practice in the public mind" and that the "tattoo was a symbolic poke-in-the-eye directed at those who were law-abiding, hard-working, family-oriented, and stable."[49] Additionally, tattoos became increasingly connected with low-class, marginal, and unconventional sub-cultural groups.[50] Bikers, convicts, whores, and gang members are only a few of the groups associated most strongly with tattooing between 1920 and 1960. Samuel Steward's (a.k.a. Phil Sparrow) ethnographic narrative of his experience as a tattoo practitioner in Chicago and Oakland from 1950 to 1965, for example, characterizes tattoo clientele under the headings: "The Fleet's In," "City Boys, Ex-Cons, and Juvenile Delinquents," and "Lovely Ladies, Tramps, Dykes, and Farm Wives."[51] Although he warns against the dangers of generalization, Steward also characterizes his clientele as largely belonging to the "disadvantaged strata of American society" who had to be handled with "inventive diplomacy" and were, except for the sailors, "not very clean."[52]

Another factor perpetuating the classification of tattoos as deviant in the mid-twentieth century was an outbreak of hepatitis in the 1950s and early 1960s. As a disease transmitted by infected needles, hepatitis became associated with the procedure of tattooing, especially because the local tattooist was usually not considered the most hygienic individual nor was he or she generally located in the best part of town.[53] The disease consequently had two significant impacts on the public perception of tattooing. Initially, it enhanced tattooing's tawdry reputation because it became a dirty, unhygienic practice associated with illness.[54] Secondly, tattooing was banned in some states as a consequence of the disease.[55] Margo Mifflin argues that these bans cemented tattooing's association with deviance because they constructed tattooing as an "outlaw" practice.[56] Mifflin's position is consistent with David Matza's argument that a ban on a practice "virtually guarantees that further disaffiliation with convention" will occur.[57]

It seems that Matza and Mifflin have a point. As recently as 2002, for instance, authors of a study published in the journal *Pediatrics* adopted without hesitation the assumption of deviance when they sought to determine the level of association between tattoos and "high-risk" behaviors in adolescents. One of the results of the study is telling with regard to the continued class evaluation of tattooing: "Tattooing was significantly associated with older age, living in a single-parent household, and lower socioeconomic status."[58] It is not that the study is wrong or inaccurate, but simply that the naturalness of assuming a connection between tattooing and deviant behavior is unquestioned. *American Family Physician*, as another example, in 1990, during the heyday of the tattoo renaissance, published an article that argued that finding a tattoo during a physical exam should alert the doctor to the possibility of an underlying psychiatric condition.[59]

The very articulation of the tattooed body as deviant, however, potentially places it in a position to distort the meaning of normalcy and deviance. The performance of deviance may disrupt the norm and, in so doing, unravel the norm's striated organization.[60] The discursive mechanism used to territorialize the tattooed body's social status and to erase its legitimacy has been the discourse of classification. Tattoo competition categories receive authority from this discourse and thus produce the tattooed body as a trace or text given for evaluation. In its reiteration of the norm, however, the competition also displaces the authority of the discourse. The new context of the tattoo convention appropriates classification and transforms it into privilege. In this venue, tattoos are no longer marks polluting the natural body; they are works of art. The tattoo is no longer the mark of the deviant mind, but the evidence of something else.

ACTING OUT. . . OR NOT?

In order to explore the something else produced through discourse appropriation requires shifting venues from Tattoo Voodoo to the 2004 New York City Tattoo Convention. The tattoo competition in New York was structurally similar to Tattoo Voodoo. The contest took place upon a stage in the ballroom and the contestants were evaluated by a panel of tattoo experts. Competition categories included best color tattoo, best back or chest tattoo, best tribal tattoo, and best overall tattooed person. Each contestant would walk onto the stage, show their tattoo to the audience, show the tattoo to the judges, and offer themselves to be photographed by the waiting photographers. The M.C. on the New York stage (unlike the Tattoo Voodoo M.C.), however, interacted with the contestants. It is in this interaction that insight into the connection between the art and the person was enacted. As I stood in the front row of the audience staring at the tattoos upon the leg of one contestant and pulling out my trusty 35mm to take a photo, I was surprised when the M.C. asked the contestant to "tell me about your tattoo." The young man began narrating a story about getting into a serious car accident and hurting his leg. He said that he had undergone fifteen different surgeries on the leg prior to getting it tattooed. The M.C. asked "did the tattoo help?" and the young man emphatically replied, "Yes!" In this instance the tattoo represents a sort of psychic healing. Instead of representing a deviant or criminal mind, the tattoo was the act of the man reclaiming ownership over the body part that was alienated through injury. He could take back his dismembered and dissociated body through the act of overinscribing it with his own interiority.

Recuperating the tattoo as positive symbolism was also the theme for a gentleman displaying a tattoo that covered much of his arm and chest. The image was of a tiger and a bear intertwined. When asked about his tattoo, the man stated that, "the tiger symbolizes me and the other half my brother." In this case, the individual tattoo has meaning for the man and within the tattoo community it goes beyond the mere fact of being tattooed. The owner of the tattoo "controls" and assigns meaning—the tattoo is an expression of some interior depth. This type of response to the question of the meaning or inspiration for tattooing in our contemporary context comes as little surprise, of course. One of the consequences of the tattoo renaissance has been a proliferation of discourses assigning personal, ritual, and/or spiritual meaning to the object and practice of tattooing. DeMello suggests, for example, that the rhetoric of the 1960s and 1970s movements of New Age, ecology, women, men, self-help, and Goddess religion led to a change in the way many middle-class Americans viewed their relationship with their bodies, the earth, and their spirituality. As a result, DeMello argues, middle-class America reinvented tattooing from a practice of desecration to one of decoration, and tattooing was recast as a means to express one's identity and spirituality (as opposed to an expression of deviance).[61] Within this discursive community, then, the classified body of the tattooed holds the seat of privilege. The culturally deviant body is judged based on norms developed by other deviants according to their internal community standards. Upper-class appearance norms are still violated by the body's display, but the process of using classification discourse to evaluate the violation as unnatural or deviant is subverted. Moreover, part of the rearticulation of the norm through classification strategies was to reinscribe the hidden interiority of the subject.

While at first glance this argument for reassigning the meaning of the external symbol through a rearticulation of motive (from deviance to distinction) appears to be a liberating move, the problem (as I indicate in chapter one) is that the structure of this discourse remains the same (is not disrupted) and therefore produces two problematic effects. Initially, the body remains the effect of the mind and therefore becomes evidence. The problem is not that the tattoo fails to become a symbol of a different, better self. The problem is of viewing the body as evidence of some essential interiority at all. As soon as the internal logic and discourse of the tattoo community stabilizes that discursive territory, the body as trace becomes controllable and finalizable. The door is open to shift the motive from spiritual transcendence back to deviance once the discursive context shifts from, for example, the tattoo community to the courtroom. Imagine, for example, the case of Elmer Pace being thrown up against a wall in Des Moines, Iowa, by police officers intent on photographing his naked torso for evidence,[62] and testimony being heard in New Mexico regarding the appeal of convicted murderers where the victim of the crime, Thomas Briggs, is being posthumously characterized as a villain partly based on his adornment and display of tattoos.[63] In short, the maintenance of the structure of the logic of tattoos as evidence on the surface of the body of some sort of essential self makes it possible for class evaluations to take place that have harmful consequences on the bodies of tattooed people.

This new discourse of tattoo as art form also produces yet another less liberating performative effect. The attempt to align tattooing with high status culture stratifies the tattoo community into high- and low-class categories. In her insightful and thorough analysis of the discursive construction of community and class by and among tattoo artists and collectors, Margo DeMello argues, for example, that

> Terms such as "biker," "sailor," or "scratcher" are used in tattoo magazines and articles on tattooing to refer to working-class tattoo practices that are said to be outmoded and are differentiated from newer practices defined as "professional" or "fine art." These are all status terms that mask class differences within the supposedly egalitarian tattoo community.[64]

Tattoo community discourse creates a class hierarchy based upon the aesthetic style of the tattoo art. Scratchers do not use or do not have access to either aesthetic training or the highest tech equipment. They work in "parlors" or "shops" and not in "studios."

In addition to this separation of scratcher work from professional or elite art, there is also an othering of the most extreme displays within the convention context. In New York, for example, the convention stage hosted a performance by contemporary sideshow performing artists Enigma and Katzen that drew mixed responses from the audience. Enigma and Katzen are Houston-based performers who are replicating late nineteenth- and early twentieth-century freak performances. Enigma's head and body are almost completely covered by a jigsaw puzzle pattern filled in with blue ink. He also has horns implanted in his shaved head. His wife, Katzen, is tattooed from head to toe with a series of black stripes meant, as her name refers, to emulate the stripes of a tiger or some other cat. In his performance, Enigma swallowed a sword and a cross. He also inserted a condom into his nose, pulled it out through his throat, and then blew it up through his nose again. He followed this by both pounding a nail and inserting a spinning drill bit into his nose. These acts are all relics of sideshow performances, primarily those popular in and after the 1920s when illustrated displays added to their performances in order to satisfy increasingly unimpressed audiences. As I watched Enigma, I listened to audience members gasp, groan, and moan in response to his nasal acrobatics. Other audience members, however, gleefully cheered and applauded. I also overheard snippets of conversation that variously evaluated the tattooing choices as "O.K. for them since they are making a living off it, but not something that I would ever consider doing," and "That's cool!" or

"Now THAT'S what I'm talkin' about!" The schism in the audience evaluation between cynical acceptance and enthusiastic endorsement demonstrates the division in the community between those embracing the deviance of the "extreme" visual and behavioral display and the more conservative response.

The effect of the effort to transform the tattooed body part from a symbol of deviance into a work of art recreates class distinctions inside the subculture, although it resists those same divisions imposed upon it from outside. Resistance through discourse *appropriation* is, therefore, limited because the shift in power relations is not comprehensive; the ideological logic of classification as a means to create status distinctions still holds some authority, and that authority is exercised within the community. The strategy of appropriating classification discourse as a means of resistance, therefore, merely reinscribes class hierarchies and stabilizes the discursive territory. In adopting mainstream cultural norms, the artistic tattooed body simultaneously replicates their logic within the tattoo community.

The act of operating within (as opposed to trying to resist through appropriation) the territorialized constraints of classification discourse, however, enabled the agentic potential of the contest entrants during Tattoo Voodoo to disrupt and stretch the boundaries of the frame because their movement emerged out of the frame. The image of the tattooed body enacted its potential for transformation, therefore, because of and through discursive territorialization and framing. Resistance to hierarchy, in this case, merely reproduced hierarchy whereas operating through discursive and visual frames actually broke the frame.

Threshold Space and the Context of Display

The events and performances taking place within the tattoo convention venue clearly kinesthetically remember freak-show and sideshow performances. All of these performances, of course, also occur within distinct spatial contexts. The site of the performance regulates the movement and evaluation of the body by becoming the frame that territorializes it as an image. In the following analysis, I argue that the spatial relations enacted at the site of tattoo convention reproduce the social marginalization of popular entertainments and entertainers. This spatial context enables the displacement of authority from the classical or pure body to the marginalized body. As a result, the spatial frame of the convention venue produces the possibility to disrupt the normative parameters of social privilege. The tattoo convention is a site where norms of behavior are upended. As Braunberger clearly states, "These events are tattoo safe-havens where the social rules governing bodies change."[65] The following analysis narrates the genealogical connections between the framed spaces of the contemporary conventions and the spatial locations of public entertainments at the turn into the twentieth century.

In 1999 and 2000, the Southern Comfort Tattoo Expo was held on the outskirts of New Orleans in the suburb of Metairie. The convention ballroom is on the lobby level in the hotel, and the entrance is clearly marked and visible from the main entrance. In order to enter the ballroom, therefore, the exposition attendee first crosses through the lobby. This lobby area, typical of most hotels in the United States, is an open area populated by hotel guests in various states of transition: people checking in and out of the hotel, waiting to meet friends, and coming in and out of the Tattoo Expo, the elevators, and the hotel bar.

The entrance to the Expo was marked by another series of gateways. The first point of entry was under a sign announcing, "Welcome To The Southern Comfort Tattoo Expo." In 1999 this sign was flanked by two life-sized cardboard pirate sentries fully regaled with eye patches, swords, and treasure chests. The ticket table was located just inside this entryway and it, too, was guarded by sentries in the form of a security guard on the right and the ticket takers on the left.

As I passed through the pirates in 1999, the soundscape of the tattoo convention beckoned me. Anxious to enter (and unaware of the price for admission), I began to walk toward the doorway to the ballroom. A t-shirt and jeans clad man with an earpiece and a walkie-talkie stood immediately in my way and grabbed for my wrist. He was looking, I soon discovered, for the neon band that would grant me entrance. I was turned back to the table I had not noticed before and, without instruction, I became rapidly aware that I was not to be let in unless I paid. Eager to make up for my mistake, I tried to attract the attention of the women in charge of the ticket table. This proved a more difficult task than I would have thought (after all, the people at Disney *want* your money, that is the whole *point*). I first tried to make eye contact, thinking that they had noticed the security guard rebuffing me. Not effective. I then performed the passive aggressive clearing-of-the-throat strategy. Not effective. Getting desperate and feeling increasingly uncomfortable both because of my own ineptitude and the seeming lack of interest on the part of the ticket vendors, I began to worry that I would never make it into the ballroom that so seductively summoned me. I decided to become more assertive and so I spoke: "Excuse me, how much does it cost to get in?" The woman on the left looked up at me from her conversation and gestured at the notice of prices on the table with a move that seemed to say, "Are you blind or just stupid?" Blushing (although I wondered why I felt as if I were at fault), I groped in my purse for money and handed her the ten-dollar, one-day admission price. She took the money and I waited for the wristband. I stood awkwardly for a moment as she returned to her conversation without giving me another thought. I slowly approached the waiting security guard, armed with the knowledge that I had paid, even if I had no wristband to prove it. As I prepared myself for a confrontation, he rewarded me with the magical band (but not until I had offered the incorrect wrist), and I was *in*!

Once my ticket was purchased, a wristband attached, and the program collected, I entered a vending area that marked the interior lobby of the Expo. Vendors sold back issues of tattoo magazines, jewelry, t-shirts, food, and alcohol. On the other side of the vendor's area was the entry to the ballroom proper where the tattoo artists, piercers, and tattoo equipment vendors were located.

The line demarcating the inside space of the convention and the outside area of the hotel lobby was very distinct. This construction of boundaries created by the series of entryways marking passage from the hotel lobby to the ballroom clearly established the lobby as exterior and the ballroom as interior. These boundaries suggest that the convention is a strongly classified space and a tightly territorialized frame. Boundaries produce areas where behaviors are deemed appropriate or inappropriate and admit or exclude bodies based upon their performance: "Difference in a strongly classified and strongly framed assemblage would be seen as deviance and a threat to the power structure. In order to minimize or counter the threat of pollution, spatial boundaries would be strong and there would be a consciousness of boundaries and spatial order."[66] In the highly striated territory, boundaries are both spatially and discursively strengthened. At the convention, tattooed bodies represented the realm of the inside and nontattooed bodies circulated along the periphery or outside. The ticket booth regulated the flow of bodies. This creation of bright-line distinctions that include and exclude specific people and behaviors is, as we shall soon see, a clear reflection of the spatial logic and social context of the sideshow. Historical authority is inventively displaced, however, to favor the marginalized as opposed to normalized body. Before returning to this point, however, I must revisit the sideshow.

The sensory experience of the convention ballroom at Tattoo Voodoo, as well as in New York and Houston, evoked the sights, sounds, and smells of carnivals in general and of sideshow tents in particular. Spatially and visually, the convention ballroom was the sideshow tent reincarnate. Sideshow tents were large, oval structures.[67] Inside the tent, "each act was curtained off from the next" and "for an extra quarter, customers were admitted to the 'Annexe,' which was touted as the most sensational or bizarre freak act of the carnival. Those freaks that did not perform often sold miniature bibles or postcards with their portraits."[68] The tattoo booths lining the ballroom walls in Metairie, Houston, and New York become the sideshow stages upon which the tattooing performance was enacted. The premier event of the convention, the "live" tattoo competition, is the "Annexe" for which convention attendees must pay an "entrance fee."

Visual cues also beg the connection between the sideshow and the convention. Perhaps the most compelling link is between the artwork lining the back walls of the booths and the banner displays advertising the sideshow. A sideshow banner line was a row of large canvas paintings strung along the length of the show's front and supposedly depicting the attractions inside.[69] Banners were large "portrait paintings" that may be lined up for as long as 140 feet.[70] Most often, these banners were vividly colorful and wildly distorted representations of the freak performers.[71] The murals of artwork lining the back walls of the convention booths produced their own banner line. Instead of advertising the body of the performer, however, these vivid displays advertised the tattooist's artistic skill. Furthermore, there is an art tradition link between banner art and tattooing. Hammer and Basker explain that some tattoo artists employed by circuses and carnivals were also responsible for producing banners.[72]

At Tattoo Voodoo, this connection was merely associative in that there were banners marking the booths. In Houston, the link was more explicit as one corner of the ballroom hosted a booth that traded in replicas of sideshow banner art. In New York in 2004, moreover, the walls of the Roseland Ballroom were decorated with what appeared to be either the actual banners or full-sized replicas of sideshow art (figure 2.3). Banners featuring illustrations of "Lobster Boy," "King of Swords," "Mentalist," and "Human Dynamo" hung high above the milling tattoo convention participants. The upstage (back)

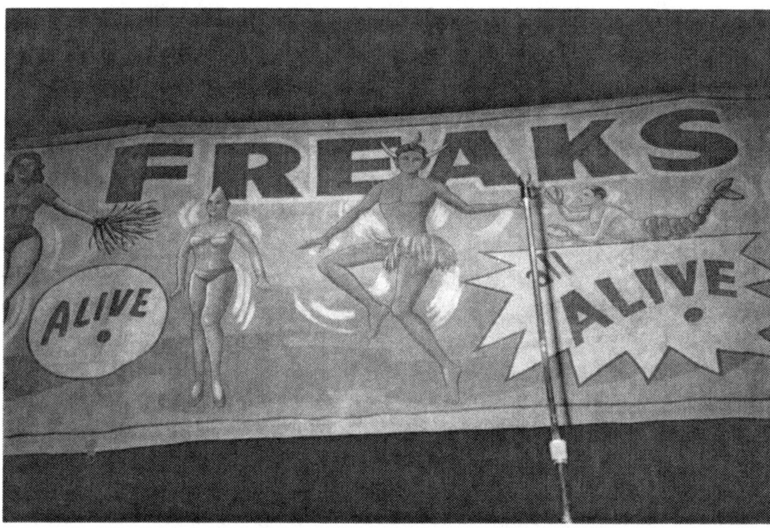

Figure 2.3 Freak Sideshow Banner. 2004 New York City Tattoo Convention.
Source: Photograph Mindy Fenske.

wall of the tattoo competition stage was framed with a banner about 24 inches long and 6 inches tall. Seven performers were featured on the banner, ranging from a picture of the "lobster boy" (with human torso, lobster claws, and a tail laying atop a bubble caption claiming "all ALIVE") to a curvaceous brunette (clad in a tight blue mini t-shirt and red short-shorts with what appear to be hands and feet of sea weed) standing next to a bubble caption announcing "ALIVE." The banner's title reads: "FREAKS."

The word "alive" or "live" on the New York banners and in the Tattoo Voodoo program further establishes the reproduction of the sideshow performance within the convention:

> The word "alive," in fact, was perhaps the most essential element of any sideshow banner. After all, these advertisements so stretched human imagination that the word "alive" was almost always added to confirm the attraction's authenticity. If the audience could be convinced that the sideshow attractions were alive, then they must also be real, although this was not always the case.[73]

When the Tattoo Voodoo convention promoters insist that the competition is "live," they invoke the aura of the sideshow.[74] The spatial and visual connection between the tattoo convention and the sideshow is, consequently, both covertly and overtly drawn by convention organizers and vendors. The tattoo convention thus depends upon and performatively revises the spatial framing of its predecessors. To understand the performative revision and potential for agency, however, requires another trip back to the sideshow.

SPATIAL OUTCASTS

The social context surrounding and producing sideshow performances at the turn of the twentieth century was highly charged with a desire of the upper classes to separate themselves from the lower classes.[75] The creation of class distinctions between high and low forms of artistic expression was one avenue through which class hierarchies were constructed and preserved. This class separation manifested itself, for instance, through the establishment of "cultural institutions" that catered to the economic and social elite:

> Efficiency combined with class bias to dictate the new cultural institutions be organized more with an eye to protecting the standard of the arts than with the theatricality and mingling of tastes which might have drawn mass audiences. The new interest in achievement and cosmopolitan excellence was not solely the product of snobbishness, as many critics of the Gilded Age have charged. The fact that many sponsors of the institutions were business tycoons meant that masterful and rather decisive temperaments directed the cultural reorganization going on.[76]

A combination of the fear of class and ethnic pollution and the guidance of a few select wealth-empowered individuals resulted in the construction of institutions that "protected" the "arts" by excluding the participation of the mass audience. The heterogeneous audiences during the antebellum era, which had so annoyed the European cultural elite, were replaced with "decorous, well-bred, and middling to wealthy" audiences who payed prohibitively high ticket prices and came attired in "full dress."[77] Moreover, the content of the new cultural institutions alienated the working-class audience because it "failed to mirror" the "needs, concerns, and anxieties" of millions of both urban and rural Americans.[78] The working class was, therefore, discursively and materially alienated from these newly created institutions of high art. As a result, popular entertainments were pushed to the social as well as spatial margins.

At the turn of the century, for example, dime museums were most often located in the entertainment districts of urban centers. These districts were located on the margins of the cultural centers of the cities. The Bowery district and Coney Island in New York, for example, were prime locations for dime museums and sideshows.[79] These areas developed into cheap entertainment districts that hosted arcades, concert saloons, and cheap melodrama as well as dime museums. The spatial location of these entertainment districts enacted the elite/upper-class attitude toward the content of the amusements. The cultural institutions of art, which the elite were constructing in their effort to create a social and economic divide between the classes, pushed popular amusements aside in terms of taste as well as marginalized them geographically. Once the institutionalization of art began after the Civil War, the dime museum lost its elite audience and was forced financially out of the center of the city. The dime museum, consequently, had no place socially, artistically, or spatially in the center of the city.[80]

The spatial marginalization of popular entertainment and the display of human oddities continued with the advent of the world's fair. At the first world's fair held in the United States in 1876, the popular amusement industry in the form of dime museums, vendors, and showmen set up *across* from the exposition's main building and *outside* of the fairgrounds in temporary structures that earned its name "Shantyville."[81] Not allowed or invited into the event that was designed to celebrate through display the best elements of the world's culture, the popular amusement industry once again took its place on the margin.

The position of popular amusements on the edge of the world's fair was reenacted once more in 1893 when Chicago was the host. Chicago's organizers, however, recognized the financial importance of the vendors and so sought to incorporate the popular amusements into the 1893 event. The financial inclusion of the Midway Plaissance was not, however, complemented with a spatial inclusion. The Midway was a 600-feet wide, mile long avenue constructed *outside* the fairgrounds proper.[82] Without the vendors, the fair could not approach financial solvency but this economic reality was not spatially sanctioned.

The economic inclusion/exclusion is complicated further by a moral evaluation of the working class. The official purpose of the Chicago fair was to celebrate the "virtues of a technological society."[83] The space representing these virtues was located in the center of the fairgrounds and was called The White City.[84] The implications of the spatial location of this shrine to technological advancement as well as the loaded racial implications of the name itself create a clear boundary between the virtuous and clean white world and the messy, immoral, and immigrant-populated working-class world. The dismal irony, of course, is that it is only through the exclusion of the working class that the upper class could construct an inclusive space for themselves.

After the 1893 Chicago Exposition, traveling carnivals began their tours across the country because the Midway Plaissance provided a forum where the various independent performers and show people converged and began to organize themselves.[85] Their marginalization and tawdry reputation traveled right along with them.[86] As these touring outdoor entertainments journeyed through the countryside, they stopped in communities and set up their temporary structures at the edge of town. Like the fringe position of "Shantyville," the carnival sits on the edge of polite society. The strongly classified space, consequently, relegates the carnival to a social and cultural margin or threshold. The combination of this physical location with the social evaluations of the performances held within it solidified the sleazy and tawdry reputation of the carnival and those associated with it.

Spatial boundaries reflected social boundaries and the early-century working class and their entertainments were subsequently shoved aside. Similarly, Tattoo Voodoo is located on the margin of the city in a suburb of New Orleans. Moreover, the Landmark Hotel is a Best Western hotel and, on the grand scale of the hotel elite, more a Holiday Inn than a Ritz-Carleton. Perhaps Tattoo Voodoo is so located because Halloween weekend is an extraordinarily popular tourist weekend in New Orleans and so finding a venue nearer the center of town is prohibitively expensive for a mainly middle-class, self-employed convention clientele.[87] Or, perhaps it is because the upscale convention venues in the French Quarter and Uptown (hotels such as the Westin, Marriott, Sheraton, and Radisson) prefer not to host a tattoo convention with all its attendant stereotypically deviant baggage.[88] In either case, Tattoo Voodoo remains at the margin. The Houston location, like Tattoo Voodoo, is also located on the edge of the city at the Park Plaza Reliant Center near the southwest corner of the 610 interstate loop around Houston.[89]

Situated immediately across the interstate is Houston's Six Flags Amusement Park. The contemporary location of the Houston Hot Rod tattoo convention on the edge of the city and next to a site of popular amusement reenacts the late nineteenth-century and early twentieth-century association between tattoo display and world's fairs and circuses. The convention is pushed, like its sideshow predecessors, to the margin.

The New York Tattoo Convention is held at the Roseland Ballroom. The Roseland was and is one of the largest commercial dance halls in New York City, capable of hosting up to 3,000 dancers or, in recent years, concert goers and tattoo convention attendees. The Roseland is located in the heart of Manhattan's theater district next door to the Virginia Theater (which was, ironically, in the middle of a run of *The Little Shop of Horrors* during the 2004 convention). At first glance, this relatively central location in an arts district apparently complicates the spatial marginalization of the tattoo convention. Despite the central location, however, the Roseland has a long association with marginality and working-class entertainment because of its history and continued status as a dance hall. During the late nineteenth and early twentieth centuries, dance halls emerged as popular entertainment for working-class populations and, like the dime museums and other popular amusement venues, were located on the edge of the entertainment district. It was the combination of location and entertainment style, moreover, that marked the dance hall as a working-class, tawdry space. The spatial location of the Roseland consequently resurrects the marginality of popular amusements and entertainments as well as their working-class heritage.

Spatially, dance halls in the late nineteenth and early twentieth century in New York were generally found in the commercial amusement districts and catered to the working class. Kathy Peiss points out:

> the large halls were usually located in the commercial amusement zones of the city, in such areas as 42nd Street and Broadway, 14th Street, and 125th Street, serving a citywide clientele. Dance palaces attracted people of all nationalities, but they appealed more to factory and office workers than to middle-class and elite amusement seekers, who flocked to Gotham's cabarets and restaurants.[90]

The original Roseland Dance Hall opened in 1919 and was located at Broadway and 53rd. The building on West 52nd opened in 1922 as an ice-skating rink and was taken over by Roseland in 1956.[91] Each location was on the edge of both the commercial entertainment district and the working-class neighborhoods. Peiss's map of dance hall locations suggests, for instance, that the Roseland was situated on the edge of an area of the greatest tenement concentration during the second decade of the twentieth century.[92] As such, the site marked the margin between the commercial entertainment districts and the working-class housing districts. Peiss's map also indicates that the areas of largest tenement concentration were along the edges of Manhattan, with the center of the island (the White City?) reserved for business and upper-class living.

The activities within dance halls also raised the ire of some social guardians of decorum and respectability. Inside the dance hall, young men and women violated norms of socially decorous performance by cavorting and gyrating under the influence of alcohol and without the benefit of chaperon. These activities so concerned Belle Israels, a social worker, that she formed the Committee on Amusements and Vacation Resources of Working Girls: "Its report disclosed that nine out of every ten girls considered dance halls their favorite resort, forty-nine out of seventy-three New York City Dance Halls sold alcohol, and twenty-two of these were attached to hotels known to be brothels."[93] Although the Roseland was not a brothel *per se*, dances with young women could be purchased for a dime-a-dance and, despite the reform efforts of social workers such as Israel, the Roseland continued to sell beer, employ dime-a-dance girls, and invite unescorted paying female patrons. As depicted in a 1937 *Life* magazine spread, women could make a fairly good living at the Roseland, pulling in an average of $45 a week.[94] Despite the claim of the magazine that most of these women were respectable secretaries by day, the cultural association between dance hall girls and prostitution remained intact.[95] Dance hall girls, in short, could make a relatively solid living, "albeit one that lurked outside the bounds of respectability."[96] Taken together, the geographic position of the dancehall and the cultural judgment of the activities taking place inside its doors firmly situate the heritage of the Roseland within the margins both spatially and culturally.

Tattoo Voodoo, the Houston Hot Rod Tattoo Convention, and the New York City Tattoo Convention sites, all provide spatial locations that resurrect and reenact the marginalization of popular entertainments in general. As such they are the territorialized frames within which the images of the tattooed bodies are located. It is within the spaces of the margins, within the striated limits, therefore, where tattoo images act.

INSIDE THE SHOW

The convention's spatial and visual citation of the interior of the sideshow replicates the logic of sideshow community within the space of the convention. Within the marginal space of the sideshow and thus in the tattoo convention, the norms for decorous performance and display are upended through both the inversion of class privilege as well as the specific mechanisms through which that inversion is made possible. Within the margin, the strategies of classification and spatial exclusion became locations for the enactment of other identities and communities. Placed in the margin of society, for instance, sideshow performers and audiences began "developing a way of life apart from the mainstream."[97] This culture privileged the freak and excluded the townsperson and was, yet again, spatially reinforced through boundary construction.

Carnival midways, for example, were divided between "front end" games and concessions and "back end" shows and entertainments.[98] Interestingly, within the culture of the carnival, the back end workers considered themselves as higher classed than the front end workers.[99] In the microcosm of the carnival, the space is still strongly classified and, once again, the periphery is the site of the lesser privileged. What is perhaps more intriguing, however, is how the classification of space within the carnival inverts the class distinctions of the outside world. From the outside, the performers in the sideshow tent were more than merely low class; they were freaks. Inside the carnival it is the freak show that occupies the central space of authority. Within the carnival entertainment industry, the most popular and often largest event on the midway, for example, was the freak show or "ten-in-one."[100] Moreover, the townspeople were not allowed to go behind the sideshow tents and enter the rear of the carnival.[101] This back-stage area where the carnival performers lived was off limits to the general public. By refusing entrance, the carnival employee used classified space to shift power relations. Outside of the carnival borders the sideshow performers and carnival workers were barred both literal and figurative entry into the physical and social spaces of the middle class and elite. Inside the world of the carnival, the outcast freak-show performers and carnival workers controlled who was allowed to enter what spaces and under what conditions.

Boundaries and spatial marginalization produce two distinct worlds: inside the sideshow/tattoo community and outside that community in the "normal" social arena. The boundaries created areas where specific behaviors were produced as acceptable based on their adherence to the social norms prevalent *inside* the borders. The lines serve as a public notice that the world on the other side invites and accepts a different set of norms. Highly classified space with strongly territorialized boundaries produced the sideshow entertainer as simultaneously sleazy and celebrated. Importantly, the territorialized lines of demarcation *enabled* the working class as much as they excluded it by constructing a space where carnival workers could enact their own class identity. Popular entertainers were outside and on the edge. Placed in the margin of polite and upper-class society, the popular entertainment industry created its own classification system as well as subverted and transgressed mainstream social customs and norms.

On/In the Threshold

The culture of the sideshow is clearly separated from mainstream culture. The easy delineation of inside/outside is, however, disrupted by the position of sideshow subculture on/in the smooth threshold. Popular amusements were pushed to the edges of the world's fair and to the outer limits of the city, and the boundary lines between those areas and the sites of cultural privilege were clearly drawn. On the other hand, these marginal places were also those through which individuals must pass in order to reach the privileged center. Vendors and popular entertainers, for example, occupied threshold space during the 1876 and 1893 world's fairs. In order to gain entrance into the exposition, the audience had to first pass through "Shantyville" and the Midway Plaissance. The audience members left their everyday lives to enter the celebration of "C"ulture exhibited by the fair but first encountered the sensual chaos of hearing hawkers and vendors calling their spiel, of witnessing human freaks on platforms and banners, of smelling and tasting a combination of meats and sweets, and of feeling the press of the crowd as they made their way through to the other side. Moreover, visitors to towns, or those living outside of the city limits, must first pass through, or at least pass by, the carnival in order to enter the city. The threshold is the gateway that everyone must pass through in order to reach the White City or the center of town.

The location of the sideshow (and its tattoo convention successor) in this threshold produces part of the power to disrupt the territorialized class structure. Bakhtin indicates the significance of threshold space in his discussion of the chronotope in "Forms of Time and Chronotope in the Novel." According to Bakhtin, a chronotope is a specific expression of how space and time interact to create environments that allow and produce specific behaviors: "Time, as it were, thickens, takes on flesh, becomes artistically visible; likewise, space becomes charged and responsive to the movements of time, plot and history."[102] The sedimentation of class structures at the line of the threshold, therefore, actually creates the discursive environment enabling boundary disruption. For instance, the threshold is a relatively smooth place containing the time of indecision and decision, crisis, and breaking points.[103] The environment of the threshold is characterized by danger, possibility, and *in*determinacy. This association suggests that within the threshold, spatial classification systems weaken and begin to deteriorate. This breakdown opens up possibilities for difference. Unlike simple boundary crossing, which has the potential to be easily reterritorialized, crossing through the gaps and fissures of threshold space is not so neatly packaged into either/or categories. Spectators passing through the threshold of the Chicago World's Fair, therefore, navigated through a place where their habitual modes of interpretation and cognitive schemas for interpreting and evaluating behavior were challenged by hybridity. They were in a world of seeming contradictions: a world that was both part of their everyday world and not, both part of the world's fair world and not, both easy to classify and defiant of classification. The upper class must pass through this space/time of instability populated by working- and lower-class amusements in order to reach the White City. The upper class must, at least for a moment, experience the threshold's destabilizing force.

Tattoo conventions performatively reproduce threshold power in their reiteration and inventive displacement of popular entertainment spatial configurations. The first threshold at Tattoo Voodoo, for instance, was the hotel's general lobby. The people populating this space represented a combination of hotel guests and employees. In other words, the hotel lobby as a threshold is a relatively weakly classified space, allowing the intermingling of people located in a variety of social positions.[104] The second threshold was the doubly guarded ticket booth area. In 1999, for example, pirates flanked the first threshold, and the ticket vendors and security guard protected the second. The pirates beckoned to those in the lobby. Their cartoon-like demeanor was reminiscent not only of the painted banners located outside of carnivals and circuses, but also of modern marketing ploys at large-scale amusement parks such as Disney Land. If this facade lured an unsuspecting tourist, however, the second-order sentries were an unsettling suggestion that perhaps the space on the other side was not meant for the general public.

Although I was certain, for example, that my money was enough to gain admission, the fact that I performed outsider status through uninformed behavior and through my body's lack of visible tattoos meant that I was unsure if I would belong to the space I was entering. This feeling persisted both years I attended the Tattoo Voodoo Expo. In 2000, the pirates were gone (a conscious effort to de-Disneyfy the event?), but the ticket table and the security guard remained. I walked confidently up to the ticket table, pausing to smile at the security guard. This year a different person dressed in jeans and a white polo shirt with "SECURITY" written across the back of his shirt turned to glance at my eagerly grinning countenance with strangely opaque white eyes. Briefly startled by the white contacts, I chuckled inwardly: "Well, what exactly did you expect, Fenske?"

The threshold was a place where codes of the outside world clashed with signifiers of the tattoo world inside. My nonvisibly tattooed body still marked me as an outsider despite my familiarity with the process of entrance. My ability to negotiate the threshold was not sufficient for me to claim insider status. My body did not display the visible evidence of my tattooed identity. For those tattooed bodies whose behavior was alienated in the lobby, therefore, the ticket booth was a table marking the "safe boundary space of the convention."[105] For the nontattooed, however, the ticket booth was the wardrobe transformed into mystical gateway: a place of both invitation and warning. As a location that canalizes behavior, therefore, the threshold produces uncertainty.

Each threshold also acted as a sort of social filter, distilling the general population of the hotel's weakly classified hotel lobby into more clearly defined cultural groups. The passage through these entryways felt very much like I was moving deeper into the tattoo world and leaving mainstream culture behind. The sentries at each entrance acted as virtual and very literal gatekeepers foreshadowing the aura of the inside. The security guard's body in the threshold, for example, represented the transition from one culture to another. These sentries, much like the ticket takers and security guards of the carnival and circus, manage the space of the threshold. They regulate the flow within and through the convention territory. The price of admission becomes a mechanism to deter the uninvited voyeurism of the nontattooed upper and middle class and/or a measure of the commitment of the potential audience member/participant. As I stood watching the tattooed bodies parade across the tattoo competition stage, for example, I was struck by the feeling that my body was out of place in this audience and this venue. Instead of feeling comfortable in my usual position of unmarked middle-class whiteness, I felt implicated because I was not (at least then) visibly tattooed. A short while ago, I was safely ensconced in the lobby of the hotel where my body was in the position of social privilege (or, at least anonymity) and the tattooed convention attendees drew stares from hotel guests. Now, forty feet and an eon later, the roles were reversed. The mainstream position of privilege is thus performatively inverted. The spatial configuration suggests a highly and strongly classified system, but the performance of privilege within this classified space is that of the culturally defined deviant body.

It was not until my trip to Houston that I had successfully adopted the style of visual display appropriate to the tattoo convention threshold. I made the effort to preregister for the Houston convention and was rewarded upon arriving with a badge with my name boldly printed and displayed. I also made a special shopping trip to purchase a halter top that would reveal most of the tattoos on my back. The rules for display within the threshold constructed by the classified space of the convention were becoming more and more familiar to me. Perhaps this is why I was rewarded by a friendly gentleman who tapped on my shoulder and requested permission to take a picture of my tattoos. In that instance, my outsider status was transformed and I felt truly in.

The marginalization and the subsequent construction of threshold space and the creation of identity and inversion of social authority occurring within the threshold successfully produces the tattoo convention as a site where mainstream social class privilege is destabilized. My experience in the threshold between the lobby of the hotel and the lobby of the Voodoo Expo was a result of the tattoo convention's reenactment of strongly classified boundaries as well as its memory of the position of sideshow subculture on the threshold. The convention, therefore, is the specific interaction between space and time that creates the environment for the tattooed body to perform differently. In the convention, as in the carnival, the position of insider privilege is the domain of the freakish body. My feeling of discomfort at not really belonging or being allowed backstage, therefore, was a resurrection of the carnival and circus bias against outsiders. The tattooed bodies and performers within the context of the twentieth-century tattoo convention reenact, therefore, the inversion of privilege constituted by the back end and front end relations of the carnival. In these ballrooms the normal becomes deviant as the ticket table becomes the threshold into the privileged and central exhibition featuring that which is usually marginalized.

The tattoo convention is clearly a performative displacement of the sideshow context. The contemporary tattoo convention's outside status, its manipulation of threshold space, and its celebration of the tattooed body and the marginalization of the nontattooed outsider are all performatively possible because of the convention's reiteration of the prior cultural performance of the sideshow. Moreover, it is the territorialized structure and the strength of the classified boundaries that empowers the display of the body within the threshold to produce tensions between inside and outside, normal and deviant. The defining spatial characteristic of the image of the tattooed body in this particular context is, therefore, indeterminacy and disruption of class privilege.

Conclusion

The extensively tattooed body on display is always marked as deviating from mainstream social norms. Efforts to align the extensively tattooed body with high status "art" have served, therefore, to reenact class oppression within the tattoo community through the efforts to endow the body with social legitimacy. Despite the tattoo convention's success in appropriating classification discourse in order to celebrate the tattooed body, however, the inversion of categories of privilege does not deconstruct the logic of categorization. Moreover, through its dependence upon the discourse of the essential and interior self, the celebration of the artistry or symbolism of the tattoo produces the continued possibility for diagnosing deviance.

The discourse of body part classification in the tattoo competition venue also isolates body parts. These parts are then evaluated according to their adherence to a set of aesthetic criteria. Legs, arms, and stomachs become flat canvases for the inscription of artwork. The tattooed body is fragmented for the purpose of display and "returns with all the problems of capturing, inferring, constituting, and presenting the whole through parts."[106] The tattoo competition context, like the scientific discourse during the late nineteenth century, controls the animation and intangibility of the body. The force of classification also rules the spatial configuration of the convention venues. Outside of the ballrooms in New Orleans, Houston, and New York, the tattooed bodies were freaks and inside they were not.

There were, however, at least two styles of disruption that occurred within the context of physical and spatial classification that offer some insight into the power of the image to work productively within territorialized discourse. The first occurred when the female contestant posed and smiled, the audience catcalled, the man jiggled his belly, and the audience laughed. As a result of this chain of events made possible through the limitations imposed by the classification of the body as art, there was a disruption of the containing force of the contest context. The woman's movement succeeded in acting differently because of the limitations imposed by the context frame. Within this classified discursive and spatial environment, movement was made possible and so was the audience's recognition of the insufficiency of the contest category as a framework for evaluating the body. It was not that her behavior failed to be contained, but rather that the containment created the possibility of her behavior. Outside of the classification discourse, her movement is simply movement. Within that frame, however, her movement is resistance.

The second disruption occurred within the context of the spatial classification and the destabilizing force of the threshold produced as a result of the strength of territorialized borders. The power of the classifying frame to produce strong borders and clear limits is the same force that results in the possibility of transformation. But for the clear distinctions between inside space and outside space, and the constraints upon behaviors that those limits produce, the threshold or the margin would not have had the capacity for disruption or destabilization. Agency in these terms is made possible because of the strength of discourse to discipline and contain. The stronger the classification, the more destabilizing the threshold.

The potential for disruption of discourse and the resulting possibilities opened up for transformation illustrated through this analysis of live tattoo performance have, as I suggested in the introductory comments to this chapter, more than a casual association with the theorization of visual image vitality and agency. Lessons learned about the ability of live performances to work within context are a powerful metaphor for the ability of the image to work within discourse. Remembering that the posture needed to envision the life of a visual image requires an "as-if" orientation, the insights about the actual performance of the bodies at the tattoo conventions can be used to understand the virtual possibilities of using performance as a way to read image. As Jennifer DeVere Brody points out, performance-based readings of art can make the image "come alive as a force in the world."[107] The live body in performance demonstrates that the ability to move within framed discourse is more than just a mere critical mirage. Arguing that strongly territorialized discursive settings are productive of agency is not, as this chapter has suggested, simply a hypothetical theoretical operation.

The following two chapters, therefore, explore these possibilities of destabilization and disruption as they occur at the site of the tattooed body on display in live performance venues in the context of analysis of advertising and photography. The next chapter specifically locates the tattooed body within the mediated frame of advertising in order to determine whether the potential to disrupt is possible when the body itself cannot move. Can the body represented in advertising break through into performance and resist reproduction?

CHAPTER THREE

ROSIE GETS A TATTOO: GENDER, FRAGMENTATION, AND ADVERTISING

ONE OF THE CONSEQUENCES OF THE TATTOO RENAISSANCE has been an increasing presence of tattooed bodies in advertising campaigns. For example, clothing and fashion labels such as Mossimo, Calvin Klein, Guess, and Polo have all utilized tattooed models in their magazine advertising.[1] The fashion industry is not alone. In 1999, for instance, Sony ran a PlayStation advertisement for a video game featuring the female animated "guitar-slinging megastar" UmJammer Lammy, depicted by a photograph of the stomach of a (presumably) female model with a tattoo around her pierced belly button proclaiming "THERE'S A NEW ROCK STAR IN TOWN."[2] The tattoo advertisement moved from print to live performance venues in 2001 when the bodies of boxers began to sport temporary tattoos.[3] In the summer of 2005, the online casino Goldenpalace.com went one step further. The casino purchased advertising space via eBay on Utah resident Kari Smith's forehead where she agreed to have the Web site address permanently tattooed for a fee of $10,000.[4] Considering advertising's role in the (re)production and communication of social values, and the trend toward taking the metaphor of the inscribed body literally in the world of tattooed advertising images, the representation of the tattooed body within that frame becomes an important site for the circulation, communication, and potential disruption of meaning.

This chapter explores the visual image of the tattooed body within the context of print advertising, gender, and the discursive strategy of fragmentation. In visual advertising discourse, like the one on the tattoo competition stage, the tattooed body is in parts. Instead of focusing on the ways in which body parts are socially "class"ified, however, this chapter enters into the image via the history and discourse of gender. I argue that body parts are strongly identified as masculine and feminine and that advertising images incorporating bodies with tattoos express and oppose this identification. I additionally suggest that the form of the advertisement represents social interaction and social space. This analysis, therefore, concentrates on the social and spatial context of the advertisement as well as on the active kinesthetic memory within the advertising image. In so doing, I argue that the tattooed body in contemporary advertisements is made meaningful through its recollection and citation of past advertisements and their social and historical contexts.

The chapter indicates that the commercialized tattooed body in the media, like its tattoo competition counterparts, rearticulates as well as reinforces norms. More importantly, the chapter locates the vitality of the image within the territorializing discursive environment of image fragmentation and recombination. Sedimented meanings adhere to individual pieces of images that then carry those meanings with them into new images and discursive fields. It is in the act of injecting the already held meaning into the new context where disruption can occur. Additionally, body part fragmentation has potential for disruption. The act of categorizing body parts as separately gendered takes the body apart and lays it out for inspection as pieces and fragments; it makes out of the active body a series of traces. The analysis suggests that the agency of the tattooed image can emerge within this frame because the act of fragmenting the body, of taking it apart, cannot simply be reversed. Once the unity of the body as a seamless object has been deconstructed, the individual parts begin to have lives of their own that struggle against losing their autonomy within a unified whole. The individual image traces/territories produced within the striated assemblage of the discourse of fragmentation can be, instead, recombined in ways that disrupt the field of, in this case, gender. Humpty Dumpty cannot be put back together again and this, I suggest, is actually a good thing. When the force of dissecting the tattooed body overcomes the possibility of unity, the images act out through the points of tension that emerge in the interplay between fragmentation and the recombinations that strive for, but will never achieve, conformity.

The specific advertisement that begins this look into the relations between gender and visual advertising images is an ad I came across in a U.S. women's magazine. In 1999 and 2000, Tampax, a division of Proctor and Gamble that manufactures a popular line of feminine hygiene products, ran a print advertising campaign in which the late 1940s World War II poster image of Rosie the Riveter was tattooed with the words "**TAMPAX** WAS

THERE" on her arm. The advertisement recalls the power of Rosie the Riveter as a cultural symbol of feminism and recodes the tension circulating between signs of masculinity and femininity in the Rosie image. It also importantly adorns her with a tattoo. The tattoo on the image of Rosie the Riveter further complicates the gender tension because the tattoo itself is also a site of gender conflict.

The choice to enter into the history of the tattooed image in advertising through the gateway of the Tampax advertisement is motivated by three considerations. Initially, the Tampax advertisement utilizes a cultural icon that gained its initial currency prior to becoming tattooed. Rosie the Riveter entered the U.S. national consciousness in the 1940s as a nontattooed body. As a result, the Tampax advertisement is important because it demonstrates how the tattoo infects the cultural meaning of the body. Secondly, Rosie is not a "real" person and so the element of motivation and individual choice is altered in the interpretive equation. Rosie did not decide to get a tattoo to express her identity or resist authority because Rosie is not a "real" woman at all. Rosie is an illustrated icon of womanhood. The meaning of the tattoo cannot, therefore, be argued to be an intentional expression of Rosie's individual autonomy or identity. To that extent, her tattooed body becomes a unique text for understanding what the tattoo as a cultural sign means when motive and psychological interpretations are highlighted as discursive attachments to the body and not as taken-for-granted truisms about identity expression and formation. Thirdly, the Tampax advertisement is the nexus of multiple historical discourses because of its appropriation of two seemingly unrelated signs: Rosie the Riveter and tattoos. The advertisement demonstrates, therefore, what happens when a cultural surrogate stands in for multiple prior representations and how those histories are related.

GENDER PERFORMATIVITY

More specifically, I am interested in how gender norms are destabilized as well as created and expressed by the symbolic performance of the image of the tattooed body. This approach to gender assumes that it is performative. Butler's now familiar description of gender performativity in *Gender Trouble* suggests that gender as a category of normative discourse is constituted in and through its signification.[5] Gender is produced through the repetition of behaviors, acts, gestures, and symbols that both are interpreted as gendered as well as provide the field for the possibility of gender. Gender is not determined by culture or biology, but rather constituted through and made possible by activity (discursive and otherwise). To turn John Sloop's phrase in *Disciplining Gender*, gender is not something that merely *is*, but something that is *done*.[6] To argue that gender is performative is, therefore, to presume that gender "is a fantasy instituted and inscribed on the surface of bodies."[7]

One of the effects of gender performativity, according to Butler, is that gender "becomes a free-floating artifice, with the consequence that man and masculine might just as easily signify a female body as a male one, and woman and feminine a male body as easily as a female one."[8] Despite the theoretical possibility of gender ambiguity, however, the unstable status of norms can be both potentially liberating as well as oppressive and exclusionary:

> Identifying with a gender under contemporary regimes of power involves identifying with a set of norms that are not realizable, and whose power and status precede the identifications by which they are insistently approximated. This "being a man" and this "being a woman" are internally unstable affairs. They are always beset by ambivalence precisely because there is a cost in every identification, the loss of some other set of identifications, the forcible approximation of a norm one never chooses, a norm that chooses us, but which we occupy, reverse, resignify to the extent that the norm fails to determine us completely.[9]

With every repetition of identification with an idealized and fantasized set of norms comes both change and loss. According to Butler, then, the failure of gender norms to fully determine identity produces gaps (inconsistencies in gender performances) that open up potential sites for resistance to the norm. As the discussion in chapter one of excess or erasure as modes of agency indicates, however, the focus in this analysis of gender will be less on the moments of excess and more on the process of repetition through which gender gets done. In other words, this chapter illustrates the discursive processes through which images and performances reiterate naturalized categories of masculine and feminine in order to expose moments when the carefully constructed camouflage of the norm mutates through repetition.[10]

The specific locations for possible mutations are, of course, diverse. This analysis focuses on four processes of gender enactment: (1) practices that produce gender divisions, including spatial configurations, (2) symbols and images from popular culture and the media that explain, express, reinforce, and sometimes oppose gender divisions, (3) components of individual identity such as clothing and presentation of self, and (4) social and organizational structures.[11] Gender is produced within multiple representations as well as through human behavior. In addition, it is inscribed and reproduced in social space as well as onto our bodies. As a result, our bodies produce themselves as gendered via their behavior, are produced as gendered in representation, and are understood as gendered based upon the gendered space within which they act.[12]

Kimmel additionally suggests that understanding this process of gender production "requires that we locate individual identity within a historically and socially specific and equally gendered place and time, and that we situate the individual within the complex matrix of our lives, our bodies, and our social and cultural environments."[13] Because gender is not a fixed identity category, it must also be historically and socially defined. Any analysis of gender, therefore, should attend to bodies and space within their specific historical and social context. The following analysis, therefore, concentrates on the processes of gender signified by images and symbols of the tattooed body located in the setting the advertisement constructs, the setting of the advertising campaign, and the larger sociohistorical setting.

The journey that I follow once I enter through the aperture of the Tampax advertisement leads down three transecting historical paths or citational chains. The first trail winds back to World War II when Norman Rockwell's Rosie the Riveter graced the front page of the *Saturday Evening Post*. The second route returns to a similar time by hearkening back to the 1950s introduction of the Marlboro Man by the Philip Morris company via the Leo Burnett advertising agency. The third avenue situates the tattooed body in advertising within the gender discourse of the late twentieth-century context through the UmJammer Lammy advertisement. These paths converge at the site of the Tampax advertisement. I argue that the Tampax advertisement is a cultural surrogate of these other representations of the body. Consequently, the advertisement's meaning as well as the gendered meaning of tattoos and the tattooed body are instances of the displaced transmission of these texts.

TAMPAX AND ROSIE THE RIVETER

The advertisement that I originally encountered appeared in the January 2000 issue of *Glamour* magazine.[14] The advertisement is, with one notable addition, an exact reproduction of a poster of Rosie the Riveter produced by the War Production Coordinating Committee in the 1940s[15] (figure 3.1). The advertisement, like the poster, depicts "a young, pretty, white woman . . . flexing her biceps while rolling up her sleeves."[16] The image is not a photograph but an illustration of a woman's torso positioned in profile before a yellow background with her head turned so it "looks directly at the viewer."[17] Above the woman's slightly tilted back head in a blue cartoon-like bubble, the phrase "We Can Do It!" is printed in white block letters. She is wearing a red scarf with white polka dots wrapped around her hair and tied in a small bow at the top of her head. Tendrils of her brown hair sneak out of the scarf at the base of the neck, temples, and forehead. Her eyebrows are finely manicured and the left is slightly cocked. Mascara-laden lashes frame her blue eyes. Her

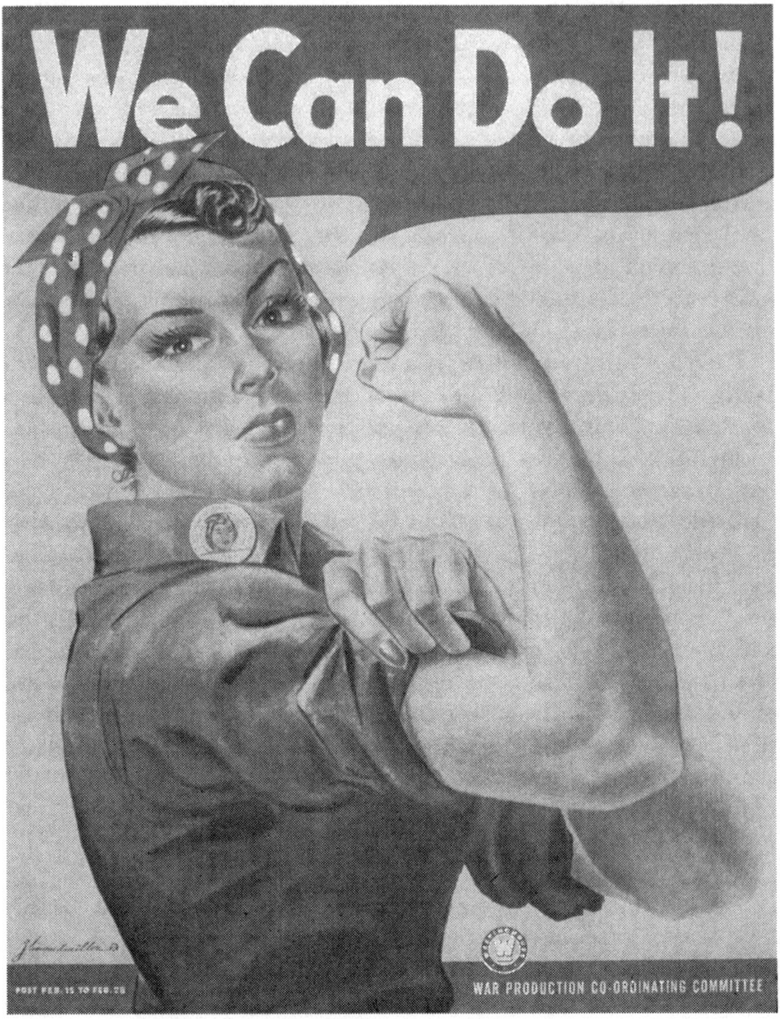

Figure 3.1 Rosie the Riveter.

Source: Courtesy National Archives, photo no. 179-WP-1563.

nose is petite, and her unsmiling, small, full lips are shaded with a light pink hue. She is wearing a blue-collared, buttoned-down, long-sleeved shirt. A round button with a black and white image of a woman's scarved head is pinned to the collar of the shirt. Her right arm is flexed and her right hand is clenched in a fist near her face. Her left hand (with short, lightly glossed, manicured finger nails) is rolling up the sleeve of the flexed arm to reveal the forearm and biceps. The sleeve of the left arm is already rolled.

The only change that the Tampax advertisement makes to the poster image is a tattoo reading "**TAMPAX** WAS THERE" imprinted upon the biceps of her right arm in black ink. The word "Tampax," in addition to being printed in a boldface, capitalized font, is surrounded by a black line that creates a rectangular frame for the word.[18]

The advertisement conspicuously recalls and (re)constructs a feminist vision of women's labor. It asks its contemporary female audience to remember the importance of female labor to the war effort. Within this context, "We Can Do It!" becomes a battle cry for the late twentieth-century woman to break through the glass ceiling with the same force with which women in the 1940s, it implies, entered the workforce. Moreover, it uses the tattoo to bridge the generational gap between the 1940s woman and her counterpart today. The image insinuates that the tattooed woman of today is the corollary to the Rosie the Riveters of World War II. The target audience of the advertisement is the generation of late twentieth-century women asserting control over their bodies by flaunting their ability to defy appearance norms. In the same way that the female laborer in World War II broke free of domestic constraints via labor, the advertisement seems to want to say, the tattooed woman of the 1990s breaks through gender constraints via control of her body (both by becoming tattooed and by using Tampax).

The meaning the advertisement desires to construct is, however, complicated by the historical authority its images recall, as well as by their situation within the context of an advertisement for tampons. The following section moves through the Tampax advertisement into Rosie's past and provides a detailed description of the war poster and its gender implications within the historical context of mid-twentieth century before returning to the Tampax advertisement's citation and revision of that history.

ROSIE'S BODY PARTS

The "original" Rosie the Riveter that the Tampax advertisement refers to is a poster based upon the Norman Rockwell print that decorated the May 29, 1943 cover of the *Saturday Evening Post*.[19] The *Post* cover depicts a robust woman dressed in blue workman's overalls sitting on a short wooden post. Her blue shirt sleeves are rolled up to reveal muscular arms. A large riveting gun rests suggestively at an angle across her substantial lap. Her bulging right arm crosses her midriff and lays atop a tin lunch box inscribed with her name in capital letters and also resting on her lap. Rosie's left arm is propped up by the elbow on the lunch box and is holding a sandwich. Rosie's short red curls are crowned with goggles and a welding mask. Her nose is pert and lips rouged. Her eyes are downcast and her face is turned away from the sandwich. Her backdrop is a rippling American flag.

In both the painting and subsequently the war poster, Rosie's female body is constructed as a masculinized worker whose purpose was to lure middle-class women into the workforce.[20] The Rosie image, consequently, mixes signs of femininity and masculinity. Importantly, those signs are inscribed upon fragmented body parts. The signs of masculinity, for instance, circulate around the body (as opposed to the head) of Rosie the Riveter. The arm in the war poster is muscular and poised defiantly with clenched fist and is revealed by the act of rolling up the sleeve of a "blue-collar" workman's shirt. The arm connotes masculine strength:

> In the sign of the arm, the contradictions implicit in the rhetoric of women and labor were exposed. These are not the bodies of middle-class women whose dainty and delicate arms, wrists, and hands performed such detailed domestic labor as needlework. These images represented the bodies of (unsexed) workers, their femininity only partially asserted.[21]

Dabakis suggests that the image of the arm is both masculine and working class. An additional implication of this statement is that the feminine image of the woman in the poster is middle class.

Moreover, masculine labor is constructed as patriotic. In the war poster, Rosie wears a blue shirt and her head is wrapped in a red and white scarf. The reference the color scheme cites is the Rockwell painting's large U.S. flag that appears suspended in midair and folded so as to suggest a waving motion. In Rockwell's version, however, the symbol of country is the backdrop and seems supportive of the Rosie figure. In the poster version, Rosie *is* the flag. In other words, the working-class masculine labor she represents merges with the symbol of the United States. Rosie is not completely masculine, however, as her femininity also participates in the poster's portrayal.

The feminine signifiers in the war poster are largely centered around Rosie's face and head. The cosmetics on the face, the perfectly manicured eyebrows, and the pouting lips with a hint of pink lipstick strongly suggest femininity. Her expression, however, is unsmiling and her cocked eyebrow challenges the viewer to deny the assertion that women "can do" the labor required from the workplace. This is hardly an expression designed to appeal to men as a sexual object. Rosie's sexual appeal is additionally subverted because her breasts are camouflaged by the masculine display of the arms. The only suggestion of her female body is the tapering of the slightly rounded line of her chest to her waist. Despite the unsmiling demeanor, however, this is the head and face of a woman who could be performing needlepoint just as easily as wielding a riveting gun. This image is unlike photographic representations of women who actually worked as unskilled industrial laborers. Photographs taken by Dorothea Lange, for example, feature women working in a California shipyard who "lacked signs of womanhood."[22] These were women who wore no makeup, dressed in workman's clothes, and, according to Dabakis, thwarted the possibility of assigning gender definitions.[23] The response of the military propaganda machine attempted to counter this image by reinscribing the working woman with domestic feminine middle-class characteristics. The Rosie of the war poster is more the image of the style and refinement of a salesperson, therefore, than that of a factory laborer.[24] In fact, taken separately from the masculine body, this woman conformed to other images of femininity constructed through magazine product advertising. She is more reminiscent of women brandishing vacuum cleaners in *Life Magazine*, for example, than of Lange's shipyard workers.[25]

Additionally, the feminine head is given voice in the poster. In a callout bubble above Rosie's head is the phrase "We Can Do It!" written in white letters on a blue background. This first appears to invert the social dominance of men because the voice (and mind) seems to provide the feminine with agency. The voice fails to complete the inversion, however, because of the mid-twentieth century interpretation of teamwork and interpersonal skills as feminine traits.[26] Femininity is located in a specific style of interaction that defaces the individual and exalts the group. When Rosie's delicate features and cosmetically correct face announce "We Can Do It!" the poster enacts the group orientation and interpersonal savvy of the feminine. The voice speaks in feminine language. Additionally, the implicit class distinctions prevent the success of the inversion. Rosie's made-up face and head signify feminine refinement while her body exemplifies working-class masculinity. The status hierarchy the poster performatively enacts, consequently, asserts the primacy of the refined feminine mind over the indelicate strength and force of the blue-collar male body. Gender and status norms are clearly performed via the kinesthetic imagination and spatial configuration of the poster. And, as Dabakis indicates, these signs coexist in an uneasy tension.[27]

This location of the feminine characteristics around the head and face is interesting because it spatially demonstrates a separation of mind and body and locates the gender division in that separation. The strong body is masculine but the mind and voice are feminine. The poster, consequently, spatially enacts this social division by highlighting the feminine characteristics of the face and head and the masculine characteristics of the arms. The feminine and masculine are uneasily connected but still separate.

The symbolic tension between masculine and feminine symbols in Rosie's image reproduced the social tension between the masculine and the feminine present during the World War II era. The need to include women in the labor force during the war years presented U.S. culture with a gender dilemma: how to maintain the line between masculine and feminine when women were performing masculinity in the workplace? The response to the tension was a strategically constructed vision of the woman at work. The basic premise underlying the vision was that female war workers did not lose their "true spirit of femininity."[28] The performance of this true femininity within the context of labor was culturally maintained through the promotion of cosmetics, the glamorization of workplace costumes, and the argument that the women were able to perform these masculine tasks only because of technological innovations.[29] The poster enacts this maintenance in the image of the working woman wearing cosmetics and glamorizing her workman's shirt with the red and white scarf.

A second conflict reproduced in the image is between domesticity and public labor as contradictory gender constructs. Evans explains that women were torn between popular images of domesticity and their growing experience of entering into the workforce.[30] The poster image rehearses the cultural confusion created when two seemingly opposed definitions of femininity coexist by combining signifiers from masculine and feminine sets of identifications. Rosie is an expert in cosmetic application and exhibits her fashion acumen by enriching the dull workman's shirt with a colorful head scarf. Conversely, her arms are muscular and she wears a decidedly unglamorous expression on her face. Rosie, therefore, articulates the ambivalence of gender identification by demonstrating Judith Butler's argument that the body both occupies and resignifies gender norms because norms always fail to fully or satisfactorily determine the body's meaning. The codes of masculinity and femininity cannot adequately express the working woman's reality during the war. The poster's construction of Rosie the Riveter reflects this inadequacy by embodying the incommensurable division between normative gender constructs. She is simultaneously masculine and feminine, coarse and socially refined.

The tension that exists within the line separating the masculine and feminine in the Rosie image produces the disruptive power of the image's performativity. As Mitchell points out in his discussion of "dialectical images," it is in the "irresolvable tensions" between "individual or generic bodies, gendered or engendered identities" where the "strange power" of image agency lies.[31] Rosie's ambivalent gender identification empowers her image to disrupt norms by performing the point at which the masculine and feminine territories collide. Neither category satisfactorily contains her experience but both are represented. The rhetorical aim of the poster to reinsert femininity into images of the working woman, therefore, merely illustrates the incommensurability of the normative constructs. Unlike the resolution of tensions discursively sanctioned by both/and subversive and reproductive positionality (as discussed in chapter one), Rosie's performative agency emerges because her masculine and feminine signifiers are specifically *not* subversive. Nor are the contrasting signifiers covered up and the differences resolved. It is the juxtaposition of purely normative signs of femininity and masculinity within the unity of the image-trace that preserves the dialectical tension between the two positions and produces the possibility of agency.

The cultural tension of dialectical gender identification that the Rosie the Riveter war poster symbolically enacts is reiterated within the contemporary context of the Tampax advertisement. The struggle with gender identification in late twentieth-century society is somewhat different than that of the war years and is enacted in discursive fields such as homophobia, sexual harassment, and the lack of adequate child care. The introduction of women into the public sphere of the workplace and the resulting struggle with gender definition that Rosie represented in the 1940s, therefore, resonates differently today. U.S. culture grapples with the dilemma between politically promoting the heterogeneity and multiplicity of gendering and the persistence of the logic of binary categorization. Despite the fact that the social and professional roles that women inhabit continue to multiply, the force of the binary distinctions between the feminine and masculine version of that role continues to exert itself within each role.[32] The masculine versus feminine tension that erupted in the mid-twentieth century has evolved into an assemblage of masculinities and femininities in tension both between and among themselves. Garrison points out, for example, that "the shift from speaking about 'women' as a unified subject to a recognition that women are not all the same, nor should they be, is something most feminists, young and not as young, take for granted in the 1990s."[33] Kimmel concurs when he notes that men are increasingly dissatisfied with traditional definitions of masculinity and are in search of new formulations that more accurately represent their everyday reality.[34] The advertisement's performative power is located, therefore, in its citation of the historical authority of the war poster within this new cultural context. The gender signifiers are rearticulated in the advertisement because the contemporary advertisement is not only located in a new cultural context where gender complexities have become more recognized but is also not a patriotic call for entry into the workforce—it is instead a pitch for tampons.

The "**TAMPAX** WAS THERE" tattoo in the advertisement implies that the women during World War II were able to accomplish masculine labor because they had the benefit of a product designed to clean up the pollution of the female body. Menstrual blood both signifies uncleanness and suggests the strong divide between men and women.[35] The advertisement insinuates that tampons emancipated women from their feminine "curse" thus enabling their effective service to country. Only by controlling the excess of the female body, the image implies, could women be valuable in the labor arena. Rosie is no longer an ambivalent image where signifiers of femininity and masculinity rest in separate and uneasy tension. She is a woman who has suppressed her femininity. The cultural reaction to Rosie's gender ambiguity as articulated in the advertisement is to interpret the signs of masculinity as evidence of successful repression of the feminine. The tension of the poster's strong dialectical divide between gender identifications is resolved by collapsing the distinction within the category of the (repressed) feminine. This is, according to Douglas, a common cultural response to anomalous figures and events.[36]

The Tampax advertisement simultaneously recalls a history of women's achievements and situates those achievements as made possible by the repression of that which is most inherently female. The advertisement's apparent desire to invoke female strength and attach that image to Tampax ultimately fails because the function of the product aligns female success with masculine values. Women only succeed, the image implies, when they simulate masculinity. The sign of Rosie and her communication of the tension between masculinity and femininity becomes, through the reterritorialization of the image within the now unifying context, a reproduction of dominant masculine privilege within the representational frame of a tampon advertisement.

The phrase "We Can Do It!" also takes on new meaning when read within the new context. Do what? Use tampons? Become masculine? Go shopping? The battle cry becomes a call for normalization. Instead of empowering the female voice with agency, the advertisement implies that the things women can do are firmly limited by their gender identity. What women can do, in other words, is placed within the context of a magazine replete with advice on how to please your man, make your marriage work, suggestions for fashion dos and don'ts, and beauty tips. In this discursive territory, women "can do" shopping, hair-coloring, plastic surgery, and marriage maintenance. On the other hand, women can also be single working mothers, support their families and otherwise succeed in a "man's" world—as long as they use tampons. In either scenario, gender ambivalence is resolved. In the former case, women are limited to acting within the confines of a singular notion of femininity. In the latter explanation, the female body is controlled in order to perform masculinity. The strategy here, according to Douglas, is the physical elimination of the contradiction.[37] Culture erases the anomaly by killing it off.[38] The only way that women "can do" men's work is if they use tampons. The question of gender tension is eliminated because the symbol of womanhood is erased as a means to gain agency. More than merely a cultural taboo, menstruation is something that prevents (masculine) action.

The choice to resituate Rosie within the contemporary community as a tattooed woman complicates matters further because of the intersection between the history of Rosie the Riveter, the history of appropriating the image of the tattoo for marketing purposes, and the history of the working-class masculinity of the tattoo. The Tampax advertisement performs another surrogacy, therefore, with its appropriation of the arm tattoo onto a female body. In order to demonstrate the meanings circulating around and performed by the tattooed version of Rosie in the contemporary advertisement, therefore, the variety of histories the tattoo transmits and displaces need to be exposed first. I will narrate one of these histories by using another popular cultural advertising representation of a tattoo-adorned body as an interpretive nexus.

THE MARLBORO MAN

The Marlboro Man of the mid-1950s provides a point of departure for a discussion of mid-twentieth-century masculinity as it is expressed via the image of the tattoo. Many of the cultural gender norms referred to by the Rosie image are likewise manifested and subverted by the 1956 Marlboro Man advertising campaign, which was also a brainchild of Leo Burnett Inc. The Marlboro Man advertisement, for example, first produced the image of the tattoo in the context of advertising. The connection between the 1956 advertisement and the contemporary Tampax advertisement begins very simply, therefore, with the Leo Burnett agency's strategic use of the tattoo as a cultural signifier. In other words, both advertisements are instances where a tattoo is inscribed upon the image or representation of the human body and not onto the body itself. That, in addition to the fact that the image of Rosie refers to the mid-twentieth century and its cultural context, connects the historicity of the Tampax advertisement to the 1950s history of the Marlboro Man. Consequently, the Marlboro Man advertising campaign is a text critical of discovering the performative significance of the Tampax advertisement's appropriation of the tattoo.

In 1955, the Philip Morris company sought to expand their consumer base that was, prior to that time, largely female.[39] Leo Burnett Inc. was subsequently employed to "engineer an image change that would increase the appeal of Marlboros."[40] The result was the advertising campaign that introduced the now familiar image of the Marlboro Man. In a 1959 interview, an executive with Leo Burnett commented that the agency "wished to show a man who, during some moment-some loose moment-got himself tattooed."[41] The early advertisements, consequently, featured "tough he-men and cowboys" who "wore tattoos on the back of their hairy hands as they lighted a cigarette."[42]

The January 21, 1957 issue of *Life* magazine features a three-page version of the Marlboro advertisement. On the first page of the black and white layout, the Marlboro Man's head, neck, and upper chest are in the frame. Wearing a light-colored cowboy hat, pinstriped buttoned-down shirt, Western style sport coat and tie, he is lighting a cigarette with a Zippo style lighter with his right hand. The back of his hand is in the foreground of the photograph and is decorated with a tattoo of the symbol of the American eagle. The heading above this image is "The Marlboro Man." The reader is invited to turn the page with the query "What's he like . . . " inscribed below the image and to the viewer's right. When the page is turned, the reader encounters a two-page layout of a series of snapshots of the Marlboro Man in outdoor settings posing most often with a horse. Each shot has a caption in quotation marks in which the Marlboro Man introduces himself.

> "I'm a rancher. Grew up in this part of the country . . . "
> "Own my own ranch . . . ride from one end to the other every day . . . "
> "I like the life a man leads out here . . . the good feeling of being your own boss."
> "Like to smoke, too. My brand's MARLBORO. In my book, it's a lot of cigarette."
> "It's got an easy-drawing filter that works just fine. Lets the real tobacco flavor come clean through, too. No huffin' or puffin' with this one."
> "And this flip-top box never mashes a one of your cigarettes. Keeps that loose tobacco out of a man's pocket, too . . . "
> "You know what they say about MARLBORO, don't you? 'You get a lot to like.' Well, that's how it is living on a ranch. You'd like that, too."

This advertisement constructs a persona for the Marlboro Man and, like the Tampax advertisement, endows the image with a voice. The narrative introduces the audience to the Marlboro Man's persona. The first three frames define the persona as a property-owning, rugged individualist who is free to roam his land. The Marlboro Man is able to "ride from one end" of his land to the other "every day." The fourth frame forges a connection between this free-spirited rancher and the act of smoking. The narrative line characterizes the Marlboro as "a lot of cigarette." The implication suggested is that bigger is always better. A man owning a large ranch and controlling his time could only be satisfied by an equally large cigarette experience. Size becomes a metaphorical expression of the masculine experience constructed by the advertisement.[43]

Size is not the only defining characteristic of the space the Marlboro Man inhabits. The ranch is additionally open, rugged, and located in nature. For example, the first snapshot in the advertisement shows him sitting tall on a horse prancing in long grass. A barbed wire fence is immediately behind the horse and the rider. A large open expanse of field stretches out in the horizon beyond the fence and a patch of woods borders the field. In the second frame, he has dismounted the horse in what appears to be a clearing in a wooded area (perhaps near the bank of a small stream because of the change in landscape from forest in the background to large boulders and rocks in the foreground). The change in scenery from field to forest implies that the wooded area bordering the first frame is also part of the property owned by the cowboy.

There is no indication that the Marlboro Man has any life whatsoever outside of his ranch; this is his sole behavioral vortex. There is also no suggestion that he suffers the constraints of urban or domestic life. For example, he wears no wedding ring. He has no ties to the private and feminine domain of home or domestic space.[44] The advertisement represents, therefore, a very specific time and space where the performance of the Marlboro Man is possible. As Bakhtin notes in his discussion of the chronotope (the intrinsic connection between time and space), nature in artistic form expresses the stretching out of time.[45] The moment of freedom in nature encapsulated in the advertising image is, for the Marlboro Man, eternal.

The vortex of the ranch is notable because it symbolizes the desire of the middle class to free itself of the boundaries of the city. City centers were no longer optimal living spaces for the growing middle class. The privilege of the center of the city as the home of the social and cultural elite was being swiftly transformed as the middle class fled urban life in droves. Residents of the suburbs were young, married, affluent enough to seek a more leisured life, and desirous of the isolated, exclusive, and larger spaces of the suburbs.[46]

The chronotope of the ranch is additionally significant because it creates a space that mid-twentieth century men could imagine as a site for the performance of masculinity. Middle-class men were in search of a new identity because the assurance that being a man guaranteed masculinity was swept out from under them as they sat in their workplace behind their desks next to women. Kimmel suggests that men's roles changed from property owner and family providers to consumers of popular culture imagery and, consequently, the concrete notion of "manhood" became the fluid norm of "masculinity":

> New definitions, for example, that indicated a historic shift in language-from manhood, the inner directed autonomous American producer, to masculinity, the set of qualities that denoted the acquisition of gender identity. While "manhood" had historically been contrasted with "childhood," to suggest that manhood meant being fully adult, responsible, and autonomous, the new opposite

of "masculinity" was "femininity," traits and attitudes associated with women not children.[47]

The distinction between masculine and feminine became increasingly difficult to manage after World War II when women entered the workforce and performed male labor. Middle-class man's role of economic provider was no longer physically enacted through hard labor and land ownership. Once male identity lost its fixed character, it needed to be actively constructed and that construct maintained. Middle-class male culture subsequently went on a search for masculinity and found it in the consumption of popular culture. For example, Gerald Early argues that the popularity of prize fighting in the mid-twentieth century was fueled by the middle-class male desire to fulfill a lack of manhood. In his discussion of the lives of Jake LaMotta and Rocky Graziano, both mid-twentieth century prize fighters, Early indicates that both came from poverty, fought viciously, and wrote successful autobiographies that were eventually adapted to film.[48]

The actors in the films, according to Early, "wound up aping on the screen the kind of mannerisms of the misunderstood, antisocial youth that Graziano cultivated in real life: Brando in the *Wild One* and *On the Waterfront*; Newman in *Somebody Up There Likes Me* and *The Left-Handed Gun*; and Dean in *Rebel without a Cause*."[49] Stereotypical performances of boxing and boxers became a way for middle-class men to reinvent notions of physical manhood.[50] In other words, behavior considered deviant in the streets became a celebration of masculinity in the ring.

The cowboy on the ranch in the Marlboro Man advertisement fulfills a similar function as the boxer in the ring for the middle-class man. The ranch is a romanticized, almost fantastical, place where men can be men. The space produces the possibility of emancipation from the constraints of urban or suburban life because of its size and its freedom from obstacles such as the office and family. It is not the fighting ring, certainly, but it is a place designed for the expression of physical exertion. The cowboy in the advertisement manages his land by riding through it daily on a horse. This is a mystical, masculine place where the men are free, women are absent, and life is good.

Entering into this complicated mix of class and gender performances is the image of the tattoo located on the back of the Marlboro Man's right hand. The hand is active, productive, and instrumentally valuable. The tattoo and the composition of the advertisement highlight the hand. The composition emphasizes the hand because the hand is the active feature of the image (it is lighting the cigarette). Moreover, this active hand is in the foreground of the image in the left corner. This is a strong position because of the natural tendency of Western readers to read from the left to right. Additionally, attention is drawn to the tattoo because the hand lighting the cigarette is the active element of the image. The tattoo is a central feature of the image that kinesthetically refers to yet another complex cultural matrix composed of issues of deviance, class, and gender relations.

Following World War II, tattoos were increasingly associated with the social underclass in the United States, largely because they were markers of nonconformity. The connection between tattoos and criminality began with scientific studies based on Caesar Lombroso's work and continued into the 1950s. For example, Steward (a.k.a. Phil Sparrow), who happened to be a personal friend of Dr. Alfred Kinsey, reports that Kinsey advised social workers and psychologists "to spend at least five full days in a tattoo shop" before getting their degree.[51] A New York policeman even went so far as to assert that a high proportion of murderers and rapists have tattoos and requested that more research be gathered investigating the nature of tattooed people.[52] The use of tattoos by prisoners and gangs to signify group membership also participated in cementing the connection between tattooing and deviance.[53] Tattooing was viewed by the middle class as a "barbaric" practice of disreputable individuals.[54]

Those who were adorned with conspicuous tattoos violated appearance norms and risked being defined as morally and socially inferior.[55] This is especially true of the hand tattoo. Hand tattoos are, at best, uncommon and are often associated with the most deviant of behaviors because they both connote a prior deviant act and cannot be hidden from the public eye. They exist in the most public of spaces, second only to facial tattoos. It is because of this very public location that some mid-twentieth century tattooists counseled their clientele against hand tattoos. For example, Steward, a former college professor and tattoo artist in Chicago during the 1950s, writes about a client named "Paul." Apparently Paul was a compulsive consumer of tattoos. Steward cautioned Paul against getting his hands tattooed and referred to the tattooist who eventually tattooed Paul's hands as "unscrupulous."[56] Even Stoney St. Clair, a former carnival tattooist who began tattooing in the 1920s, regrets his hand tattoos because they make him "look like a show-off."[57] During the mid-century, therefore, hand tattoos were considered highly public space and signified either excessive display within the tattoo community or deviance among the general public. The tattoo on the hand of the Marlboro Man, therefore, opposed the norms of male tattoo display in everyday life performance while simultaneously authorizing the Marlboro Man's performance as an icon of masculinity.

Tattoos were associated not only with lower classes and deviant behavior but with masculinity as well. As Steward points out, "A tattoo allies its wearer-at least in his own mind-with the tough, the real, the macho. It represents the submerged desires of a large part of the male population. It leaves no doubt as to one's masculinity-according to the truckdriver [*sic*] with the bevy of girls decorating his pubic region."[58] As Steward's ironic tone illustrates, there was a very specific, low-class brand of masculinity that becoming tattooed performed.

Steward's evaluation is largely the result of his unique perspective on tattooing. Although Steward was a tattooist for eighteen years, he entered the profession only after becoming disillusioned with his twenty-year career as a professor of English. Additionally, Steward's book is the product of a journal he began writing at Kinsey's request. The purpose of the journal, which was sent to the Kinsey Institute in installments, was to transcribe Steward's impressions about the potential sexual motivations for getting tattooed.[59] In other words, Steward's ironic tone expresses his perspective not only as a tattooist, but also as an ivory tower ethnographer charged with a very specific and assumptive directive. The masculinity that tattooing performs, from the perspective of a college professor turned skidrow tattoo artist, is an underclass version of sexual possession and control.[60]

The masculinity of the tattoo in the Marlboro advertisement is also derived from the citation of the authority of turn-of-the-century illustrated men. Some of the early freak-show performers were former sailors who had acquired their tattoos on sea voyages and, upon returning home, found that people would pay to gaze upon their tattooed bodies.[61] The first Western bodies associated with tattoos, then, were male bodies. The association between military service and tattooing continued in World Wars I and II as military men became inscribed with symbols of their military unit or more general images of patriotism such as the American flag. The American eagle on the cowboy's hand quotes this history, even though the narrative in the advertisement does not mention a military career. As a result, the Marlboro cowboy's "loose moment" invites the reader to perform the potential narrative of a young sailor or soldier on a drunken leave meandering into a tattoo parlor with his buddies.

The tolerance of pain required for tattooing also participates in its masculine status. The vision of the tattoo as a performance of pain is present in the narratives surrounding the exhibition of the first illustrated men. The stories told by and about these human picture galleries were those of capture and torture at the hands of primitive cultures. Former circus clown Robert Sherwood, for example, offers the following description of the process of tattooing Barnum's Captain Costentenus (figure 3.2): "Costentenus fell a victim of the tattooing needle through his Christian faith. Surely the Christian martyrs in the Roman arenas suffered no more than he. The pain must have been terrible, although not affecting the brain."[62] Likening the pain of tattooing to that of death by lion attack may seem extreme by today's standards, but prior to the invention of the electric tattoo machine in the 1880s, tattooing was a process of hammering the ink into the skin.[63] From the earliest contexts of

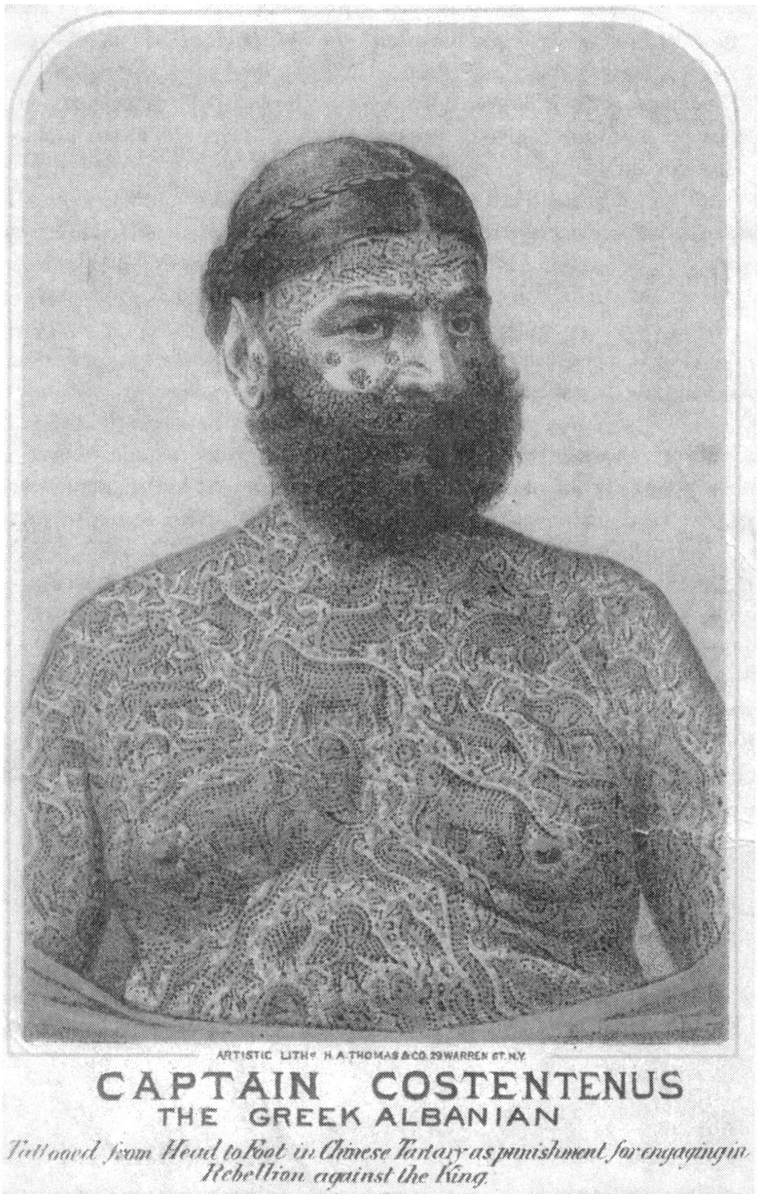

Figure 3.2 Captain Costentenus. Tattooed Man.

Source: Ronald G. Becker Collection of Charles Eisenmann Photographs. Special Collections Research Center, Syracuse University Library.

display, therefore, the tattooing process and the tattoo image have been linked to pain and manhood. This connection continues into the twentieth century because, as Kim Hewitt explains, the endurance of pain is "often considered crucial to gender construction and demonstration of toughness."[64] The spectacle of the Marlboro Man in the mid-twentieth century (as well as contemporary tattoo displays) consequently retains the kinesthetic memory of Costentenus's pain.

The Tampax Rosie the Riveter advertisement cites the complex history of the tattoo partly constructed and performed in the Marlboro Man advertising campaign. The Tampax advertisement cites the authority of the Marlboro Man advertisements of the 1950s not only because they are both mid-twentieth century bodies with tattoos in advertisements, but also because of the placement of the tattoo on the arm of Rosie and its implications for the masculinity of tattoo display. Consequently, read within this discursive construction, the tattoo on Rosie's arm in the Tampax advertisement produces the image of masculinity because of the masculine connotation of the tattoo in general, as well as because of the intersection between the tattoo and the masculine and feminine signifiers present in the original Rosie the Riveter poster to which the advertisement refers.

Initially, the tattoo is placed on the highly masculinized arm (specifically the biceps) of Rosie. According to Clinton Sanders' research, "55 percent of the questionnaire respondents received their first tattoo on the arm or hand (71 percent of the males and 19 percent of the females). Eighty-one percent (twenty-two) of the men's tattoos were on their arms."[65] Sanders's concludes that men are more likely to choose arm tattoos because of technical factors such as they are easier for the artist and because arm work is less painful.[66] Regardless of the motivation, however, the presence of tattoos on male arms and not female arms has masculinized the arm tattoo. Additionally, the image celebrates a particular version of working-class masculinity and manual labor.

Furthermore, because of the separation between the head and the body as feminine and masculine respectively on the World War II poster, the placement of the tattoo in a masculine region in the advertisement is merely an additional symbol of masculinity. It does not confront gender norms of display and control because it is not necessarily on the body of a woman. In other words, the tattoo on the arm cites the authority of the masculine codes of the body in the poster and the masculinity of tattoos in general and of arm tattoos in particular.

Finally, the context of the advertisement reterritorializes and resolves the masculine/feminine tension. As my previous discussion of context suggested, the advertisement negates the gender tension because of the implications of the product advertised and the codes of the magazine. Unlike in the poster, which maintained gender tension by clearly separating the body and head, the advertisement begins to fuse the two by valuing the masculine over the feminine. The masculine tattoo, therefore, merely reasserts that the body of the woman can be successful in the workplace as long as she simulates masculinity. The body and the head of the poster remain separately coded as masculine and feminine, but the tension produced in the division between the head and body in the war poster is subdued by context. The tattoo merely participates in this process as a symbol of yet another way the female body must be emancipated from her femininity. In this way, the agency that emerged because of the point of contact between the territorialized notions of masculine and feminine in the war poster is subsumed by the new arrangement of history and discourse in the advertisement. The motion of gender territorialization in the advertisement captures and closes down agency because the rift between masculine and feminine is mended by the force of context and the advertised product. The integrity of the gendered body parts, in other words, is weakened by context.

UMJAMMER LAMMY

There is, however, one more history to be told before drawing any final conclusions about gender and the performative agency of the Tampax advertisement. Where there is tattoo masculinity, the story goes, there is also tattoo femininity. The advertisement for Sony PlayStation's UmJammer Lammy (a video game) provides an entry point for the discussion of tattoos as a feminine code. My purpose here is to bring the discourses of advertising, gender, and the tattooed body into the late twentieth-century context with a photographic representation of a female tattooed body before returning to the illustrated form of Rosie the Riveter. The UmJammer Lammy advertisement functions as an entryway into the discourses of femininity surrounding female tattooed bodies throughout the century. Unlike Rosie, the female form in the Sony advertisement is a photograph of an already-tattooed woman's body and so is an example of how the female tattooed body (as opposed to an illustrated image of a female body) is constructed and represented in contemporary advertising. This is important because the Tampax advertisement is a surrogate not only of the "original" Rosie the Riveter (who was a surrogate of the 1940s female laborer), and of the image of the mid-twentieth century tattooed body in advertising, but also of the image of the tattooed woman in the1990s. Surrogacy is not, in other words, a process of transformation that is caught in linear time. Assemblages move multi-directionally and at various levels.

The September 1999 issue of *Ray Gun* magazine carried the UmJammer Lammy advertisement. *Ray Gun* was an alternative rock music magazine published in Santa Monica, California. The September 1999 issue featured commentary about Seattle-based rock group Soundgarden in addition to a column written by rock music bad boy Kid Rock. The magazine also ran an article about the British girl group 21st Century Girls. The article's style and content indicate the general tenor of the magazine: "21st Century Girls (known previously as Teenage Rampage, until it was discovered that a porn website uses that name) is a blend of punk rock and manufactured teen pop—a hybrid that is unsettling, if not just really fucking weird. It's as if Joan Jett was run through the production machine behind Brittany Spears."[67] The magazine was an in-your-face hip/alternative perspective on music, film, and fashion catering to the image-conscious, techno-savvy teen and twenty-something of the late twentieth century.

The UmJammer Lammy advertisement appears on pages six and seven of the twenty-three- page advertising onslaught at the beginning of the issue. The advertisement features the torso (from just below the breast line to the hips) of a woman wearing jeans and a red cropped t-shirt top that reveals her stomach. The shirt is the same costume that her animated compadre wears in an illustrated box at the bottom of the advertisement. The woman's navel is pierced and surrounded by a black and gray tattoo design reading "THERE'S A NEW ROCK STAR IN TOWN." The written text of the advertisement indicates that "Lammy" is "the guitar-slinging megastar of UmJammer Lammy, the new rock video game. Problem is, she's late for her big gig with Milk Can, her all-grrrl band. Save the day by jamming to rock, punk, pop, and other killer riffs."[68]

The advertisement obviously cites the late twentieth-century discourse of "grrrl" power that, as indicated by the written text, is wrapped up in female rock star celebrity. According to Gilbert and Kile, the term "grrrl" was coined by punk rocker Kathleen Hanna and is "a spontaneous young-feminist reclamation of the word 'girl'" intended to "recall the naughty, confident and curious ten-year-olds we were before society made it clear it was time to stop being loud and playing with boys."[69] Garrison extends this notion when she argues that young women primarily linked with punk subculture in the United States and Great Britain have identified with the term and mobilized a grassroots feminist campaign. Garrison suggests that these young women claim "feminist agency for themselves" by making use of print and visual media, music genres, technologies, nonviolent action, shock tactics, and the Internet. Moreover, a significant sign associated with "grrrl" power within the context of the advertisement is the tattoo located directly in the center of the image. As the discussion of Irene Woodward in chapter two suggests, American women have been adorning (and displaying) their bodies with tattoos for over a century. The history of inscribing tattoos upon female bodies is, therefore, part of the cultural authority upon which the performative force of this as well as the Tampax, advertisement depends.

The history of the tattooed woman is one built around the eroticism of exposed female body parts. The illustrated women who performed in freak shows at the turn of the century were eroticized on the stage by violating Victorian norms of appropriate levels of physical exposure. The illustrated woman's performance was, in some ways, culturally received as a unique version of a peep show or an all-girl review:

> One reason for their appeal was that in order to show their tattoos, they had to expose parts of their bodies—their legs and thighs—which under any other circumstances would have been lewd if not illegal. This gave showmen a way of sliding a little bawdiness into the freak show tent, an act that both paralleled and facilitated the incorporation of hoochy koochy shows in the popular amusement industry.[70]

The women who performed in such acts were consequently judged, even by some of their cohorts, for their presumably lewd behavior. Illustrated woman Betty Broadbent, for example, took pains to distinguish herself from the "carnival floozies with one or two tattoos who would bump and grind."[71] As a result of their flesh exposure, these women "sacrificed social respectability."[72] The conventions of the illustrated person's performance required that the illustrated woman reveal her body; even though it was covered with tattoos, the very public act of being undressed was (and still is) erotic. This perception was reinforced because tattooed women covered themselves off stage. Mifflin suggests economic considerations motivated this move: "offstage, they covered themselves both to protect their work from the sun's damaging rays and to ensure that only paying customers took in the show."[73] It is tempting to take this explanation at face value. It makes sense that women who were willing to undress on stage would not suffer the social pressure to maintain decorum once off stage. On the other hand, the body of the illustrated woman on stage was in a heightened communicative frame. She was staged in a "highly formalized and aesthetically elaborated performance form" with a specific set of rules articulated by the conventions of display.[74] The conventions required that the women wear very little clothing. Once the tattooed woman exited the stage and entered a new communicative frame, therefore, the rules for exposure changed. The desire to cover her body may have been to preserve her work and guarantee her employment, but the social ramifications of that act suggest that tattoo exposure was inappropriate in everyday interaction. If you wanted to see the body of the illustrated woman, you had to pay for it. Tattoos became a sign of sexual deviance because, like prostitutes and strippers, tattooed women's exposed bodies were purchasable commodities.

The association between women with tattoos and sexual desire continued into the mid-century. In an era where tattooed women no longer graced sideshow stages, however, the matter of public and personal space became transferred to the location of the tattoo on the female body. Steward's account of his tattooing business in Chicago between 1950 and 1965, for example, suggests that his male clientele viewed tattoos on hidden female body parts as highly erotic:

> Earl, a youthful gang member in Chicago, heavily tattooed, and looking fierce as a young Valentino with side-burns down to his jawline, one evening brought in his young wife and demanded that his name be put high on her leg. She was too embarrassed to lower her levis (*sic*) while I was present so I stepped outside. When I went back she had put on Earl's shirt and sat with her fat white thighs pressed close together. She yelled and squirmed a lot while I put on a small rose with Earl's name beneath. When it was done she stood up and Earl suddenly grabbed her from behind, naked lust in his eyes and gestures. If they had been alone he would have topped her there and then.[75]

Nor was this an isolated occasion in Steward's experience: "Over the years nearly a hundred men brought their women to have the man's name inscribed above the celestial gate, or on the gluteus maximus. After I had moved to California and been discovered by the Hell's Angels, I marked their women's breasts many time with 'Property of' followed by the member's gang-name."[76]

Steward's narrative must always be read with his subjectivity and dual persona as tattooist and sex researcher in mind. Other female bodies were tattooed in the mid-twentieth century and not all because their men wanted them to.[77] What is interesting about the masculine subjectivity of his comment, however, is that it cites the turn-of-the-century history of the politics of body part display and concealment as an authorizing discourse for interpreting the mid-twentieth century female tattooed body. Moreover, the association between women with tattoos and sexuality, specifically sexual promiscuity, was clearly an image that permeated the mid-twentieth century American consciousness. The most compelling illustration of this association came when "a guilty verdict against a Boston rapist was overturned after a small butterfly was discovered on the leg of the victim."[78]

The implications of hidden versus exposed tattoos in terms of sexuality and the eroticized female body continue to resonate in the late twentieth century. One of the characteristics of the tattoo renaissance has been an increased number of middle-class women adorning themselves with tattoos. The key, however, is that these designs tend to be coded as feminine because of their size, design, and location. Braunberger suggests that between the 1970s and 1990s, for example, the most popular designs for women tended to be "feminine" styles such as small hearts, flowers, and butterflies.[79] Moreover, the placement of these designs tends to be on hidden body parts or body parts that have become, for the female body, sexually coded. For example, when lecturing about tattoos as a form of visual communication in an introductory communication course at a large university in the southern United States, I led a discussion about gender. Without exception, students suggest that "feminine" tattoos are small and generally located on the breasts, stomach, lower back, ankles, and shoulder blades. Part of the explanation for this could be the late-1990s fashion trend for cropped tops revealing women's stomachs and lower backs. This trend merely confirms, however, that the stomach and lower back are highly sexualized and eroticized female body parts. Moreover, these areas are only visible when and if a woman chooses (or is forced) to reveal them. The implication is that tattooed flesh is private space that, when strategically revealed, is erotic.

That sexuality has, as the UmJammer advertisement insinuates, also been recently transfigured into empowerment.[80] "Grrrls" are not sweet and innocent, nor are they pseudomen. "Grrrls" rejoice in their female sexuality because they are in control. Part of this control is evident in the reclamation of the word "girl" from its pejorative subtext by "putting the growl back in our pussycat throats."[81] Specifically, female actors such as Angelina Jolie and Drew Barrymore flaunt their tattoos. Comedian-turned-fashion-critic Joan Rivers even comments when Jolie wears clothing that *covers* her tattoos. Girl-power music groups such as the Dixie Chicks wear tattoos on their exposed ankles. Phoebe, the sexy young witch of Aaron Spelling's TV show *Charmed* played by Alyssa Milano, occasionally (depending on hair style), reveals a tattoo that trails down the back of her neck. Even Cher, the 1960s "I Got You Babe" sensation turned 1990s diva, sports multiple tattoos. These women, it could be argued, have become the illustrated women of modern popular entertainment. Instead of being ostracized for their tattoos, however, they are modeled by a generation of young women. The tattoo on these women appears to be an effort on the part of celebrities to appropriate the deviance of tattoo art. Tattoos therefore become faddish signifiers of countercultural status, rebellion, and sexuality, drawing their meanings from the historical interpretations of the female tattooed body and recoding them on a contemporary celebrity body. Mifflin argues that this move successfully runs "roughshod over socially-sanctioned visions of femininity, flouting conventional expectations as well as those of some feminist factions."[82] Braunberger additionally articulates this power of the visibility of the female tattooed form in terms of monster beauty. She asserts that women have found "transgressive opportunities" in tattoos that defy cultural proprietorship of the female body.[83] Tattoos on celebrity females, therefore, appropriate the deviance of the lower-class and masculine association. The power of their celebrity, however, alters that meaning by transforming deviance into power.

The UmJammer advertisement participates in this discourse with its desire to associate the animated video-game character with an attractive, sexual, tattooed, female body. The animated character, as a representative of "grrrls", is a "mega star" in a rock band. "Lammy" has asserted her position in the traditionally male-dominated rock music industry. With the tattoo, she has also asserted control over her body and sexuality and has broken convention by staging the public display of that control. The problem with this scenario is that the body in the advertisement simultaneously conforms to norms of feminine tattoo display. She is adorned with a stomach tattoo that could just as easily be concealed as it is revealed. Moreover, this representation of "grrrls" is designed to sell a Sony video game. As a result, it participates in what Garrison calls the watering down of the political content of the movement by commodifying the "grrrl" image. The subcultural punk feminist is reduced to an animated video-game character who, as the advertisement's text indicates, "is late for her big gig" and needs to be "saved" by the video-game player. This representation of a female character who is stereotypically late and in need of help effectively negates the empowering rhetoric of the 1990s feminist punk subculture.[84]

It is within this contemporary cultural context that the Tampax advertisement is displaced. Rosie's tattoo, on the one hand, declares that she, as a strong and powerful woman, can easily and publicly adopt a masculine image. The Rosie image, like the punk "grrrl," is liberated by "appropriating the objects, spaces, and aesthetics of a culture generally dominated and determined by men and male issues."[85] At this level, the Tampax advertisement symbolizes the effort of women to break free of feminine norms and regain control over their body parts. In fact, the image is potentially even more powerful than the UmJammer depiction because it does not conform to the norms of feminine tattoo display. Just as Disneyland, in Baudrillard's estimation, represents an American town as a hyperreal amalgamation of images celebrating values and ideals that are only an imaginative cultural construction, Rosie the Riveter represents a synthesis of an image of a 1940s female laborer with a 1990s tattooed female actor or rock star.[86] As a pop culture icon and celebrity, therefore, Rosie performs a version of World War II–girl power.

On the other hand, however, "grrrl" power is yet again commodified. The representation successfully reinvents the female tattooed form, therefore, when it appropriates a masculine symbol upon a female body within the cultural context of female empowerment. It fails because the context of mainstream media and advertising simply cannot evoke the "grrrl" ethic without normalizing consequences.

CONCLUSION

It is tempting at this point to declare that one out of three is not bad. At least read within the context of female empowerment the tattooed female body resists normative discourse. The advertisement does gain some performative authority from the rearticulation of tattoos as a symbol of sexual deviance into a symbol of female empowerment despite the fact that the advertisement is for tampons and that it represents an already masculine coded body overinscribed with an additional symbol of masculinity. Unfortunately, however, the rhetoric of female empowerment cannot be separated from the territorialized representational context. The inscription of the tattoo upon the image of Rosie the Riveter also reinforces the gender coding of the tattooed body by placing a masculine tattoo on a body already understood as masculine as well as by placing that image within the representational context of a tampon advertisement in a women's magazine. The effect of female empowerment is, like the gender tension of the war poster, muted by the controlling force of the territorialized context. Females are empowered because of tampons and their ability to perform socially coded feminine behaviors such as shopping and pleasing their men.

In addition, the advertisement fails to construct a tattooed Rosie as a productive image of female empowerment because Rosie is not a real woman but an illustrated surrogate for womanhood. At this level, I question the efficacy of empowerment discourse for women who are neither illustrations nor celebrities. In other words, these images do not reflect the reality of the tattooed woman who "must face other women still trapped" by the spell of culture's claim upon their bodies.[87] Even UmJammer Lammy fails to break this spell because she is a video-game character and, like Rosie, not "real." The effort to cut across and through the territorialized norms of gender through subject-driven empowerment, therefore, merely highlights the inability of "real" women to break free of the territory.

Despite its shortcomings, however, the Tampax advertisement and the discursive history that it relies upon and repeats do suggest some potential for agency. A lesson learned from the potential of space realized at the tattoo convention can, for example, be translated in terms of representational context. The power of the weakly classified threshold space of the tattoo convention implies that when the strength of boundaries weakens, the potential for difference enters discursive and physical space. Similarly, weakened territorialized boundaries may offer the potential for the image of the tattooed body to resist reproduction through performative repetition. The Tampax advertisement is strictly bound within the context of the tampon advertisement as well as the woman's fashion magazine. These are strongly coded frames or, in terms of the spatial analogy, strongly territorialized classified space. Displacing the image into a representational frame with weaker codes, therefore, may open up the potential for disruption.

The second option is to displace a strongly coded image such as Rosie into a similarly strongly coded representational frame that opposes the codes implied by the image. For example, displacing the Rosie image into a representational frame like a men's magazine such as *Men's Health* or a tattoo publication such as *Skin and Ink* may produce a very different read of the image. In the former case, the advertisement would be contextualized by a magazine featuring articles on the maintenance of the male body and how to please one's woman in bed. The image of Rosie with a tattoo may then be read as an example of the male inability to grasp the totality of female experience. On the other hand, it might be perceived as yet another female invasion into masculine territory. In either case, the tensions are exposed rather than elided. Within the *Skin and Ink* frame, the Tampax advertising may even take on a parodic quality because it would be read against images produced from within the tattoo community. The readership may see the advertisement as a superficial and humorous effort by the mainstream community to appropriate images from their community only to end up revealing their laughable inability to do so. The image consequently becomes a parody. Moreover, positioning the tattooed Rosie against a photograph of a heavily tattooed body on a contest stage may lead to a questioning of the politics of the construction of Rosie as a tattooed woman as well as of the appropriation of the tattoo image to sell tampons. Within these potential new contexts, the meaning of the "original" Rosie image is empowered *because* it is so strongly coded. Rather than drowning out oppositional readings, therefore, the force of the authority the image recalls is parodically confronted by the strength of the discourse authorized by the new context. The possibility for agency increases with the strength of the striated space.[88]

INFECTION

What these options fail to do, however, is to suggest that the image has agency within discursive repetition alone. That is, the aforementioned possibilities for agency rely upon an external agent manipulating the visual imagery in order to effect deterritorialization by weakening the context, or by acting to reterritorialize and recontextualize the image. Rosie and the tattoo, however, also teach a lesson about the power of the image itself. That lesson has to do with the potential of individual image traces to hold on to their meanings as they shift into different visual and discursive contexts. Both tattoos and Rosie are highly saturated with symbolic cultural meaning. Moreover, both have absolutely demonstrated their individual ability to reproduce and proliferate. Just by performing a quick and easy "Image" search on Google for

"Rosie the Riveter" I came up with hundreds of hits featuring Rosie on posters, lunch boxes, pins, stamps, and t-shirts.[89] Colleagues and students, knowing my particular fascination with Rosie, constantly bring me examples of how her image has been used to, for example, Save Martha: a poster marketed by savemartha.com to acquit, and now pardon, Martha Stewart. Rosie herself has even become a tattoo.[90] These are images that have certainly demonstrated that they have their own legs. What might happen, then, if tattoos and Rosie run into each other differently than in service to a tampon-marketing campaign? In short, the particular collision of the tattoo and Rosie in the Tampax ad was not entirely disruptive of the strongly classified territories of feminine and masculine normative discourse, but that does not necessarily mean that other interactions between the two might not be more productive. What can be learned about agency in this example, then, is about infecting one image-trace with another. The Tampax advertisement actually hints at this possibility by injecting the strongly coded tattoo image into Rosie the Riveter. The failure in terms of agency here, however, is that the subject of the tattoo (tampons) is additionally saturated symbolically and prevents the possibilities of the tattoo infecting the Rosie icon in positively reproductive ways.

There is another example of the tattoo and Rosie relation that hints at a more productive relation. The Back Up Your Birth Control emergency contraceptive (EC) campaign coordinated by the Institute for Reproductive Health Access has chosen Rosie as their poster woman.[91] This image of Rosie is available for purchase in the form of tattoos, postcards, posters, stickers, and buttons.[92] In the campaign, the war poster version of Rosie is headed with the call to "Back Up Your Birth Control with EC." The heading is in all capital letters but the font of "Back Up" and "EC" is larger and in bold red. Rosie is in her regular pose before a yellow background and is encircled around her waist with a red ribbon that reads "EMERGENCY CONTRACEPTION" in white capital letters. The text below the image admonishes the (female) reader to "Be prepared to protect yourself in case: the condom breaks, you forgot to take your pill, or, you had sex when you didn't want or plan to." This is followed by instructions on how and when to take EC and where to call for more information. The commas in the text are represented in the poster with an image of an arrow-pierced heart with the letters "EC" in the center—an image that is, importantly, also tattooed upon the biceps of Rosie's flexed arm. On the arm, the heart and arrow image is in black line and the letters in a bright red font.

The particular tattoo etched upon Rosie's arm in this image carries with it symbolic associations intersecting issues of religion, gender, and class. The arrow-pierced heart has a long history in tattooing and is, in one way, a watered-down version of the Christian image of the sacred heart.[93] In Catholic doctrine, the heart of Christ wounded by a sword or lance symbolizes Jesus' devotion to humanity. The wound then becomes the representation of the injury or pain of that love. When translated into tattoo imagery, the religious overtones of the image survive by communicating the strength or intensity of the passion rather than the source. In short, this is not Jesus' love, but it sure is strong. In terms of gender and class, the pierced heart is most often associated with working-class masculine expressions of love. For instance, early twentieth-century tattoo artists, whose clientele was primarily "sailors, Bowery bums, and other stereotypical early tattooed white guys," advertised and sold pierced heart tattoos.[94] For these tattoo clients, pierced hearts were often accessorized with the names of sweethearts or, alternately, with "MOM" or "mother." Within the EC campaign, this is a low-brow masculine image encoded with the theme of masculine "true love" placed upon Rosie's masculine arm. Instead of love for girlfriend, wife, or mother, however, the pierced heart is infected with the symbol for emergency contraception. This is a powerful surrogation in terms of gender conflict because the docile female object (girlfriend, mom) pierced by masculine love is replaced by a symbol (EC) suggesting female agency in the act and consequences of sex. Religious overtones also infect the image as the Christian symbolism of the Sacred Heart collides with the controversy surrounding emergency contraception.

In this case, the gender tension of the original war poster is replicated within the tattoo itself. The relational gender tensions and conflicts infused in the controversy over emergency contraception, as illustrated in the case of individual pharmacists refusing to dispense the medication, are repeated in the process of the relations between the histories and discourses of the pierced heart, tattooing, and Rosie herself. The designers of this campaign were particularly savvy in this choice, of course, but the force of the image has as much (if not more) to do with the reproducibility of Rosie, of the pierced heart, and of the tattoo. Moreover, the capacity of the composite image to host these tensions and conflicts is produced because the gender discourses are so intensely territorialized at the sites of these individual image-traces. As Rosie, the heart, and the tattoo are reproduced, in other words, their symbolic integrity is maintained. These images and their incessant reproducibility show how a version of viral infection can move within images to evoke tense disruptions through normative discourses designed to capture and territorialize.

Recombination

It should also not be lost on us that the practices of representation discussed throughout this chapter (and book) all implicate human bodies. Images of tattooed women and men are, as the prior analysis illustrates, dissected into fragmented body parts that accumulate and circulate gendered meaning. It is within the process of fragmentation and, most importantly, the possibilities of insufficient recombination where another capacity for agency emerges. As I hinted earlier in this chapter, perhaps it is not such a bad thing that Humpty Dumpty was not put completely back together again. In order to explain this, a brief discussion of the meanings traditionally attributed to body fragmentation is in order.

This practice of fragmenting the body has a long history dating back to (at least) the sixteenth century.[95] Fragmentation in the earliest sense represented mourning for a lost totality and/or a strategy to diffuse the power of that totality by dissecting it into manageable parts—a representational version of divide and conquer, as it were. The representations of individual body parts subsequently stood in for a missing whole. The hand represents the sailor, for instance, and the belly the reproductive woman. Wegenstein argues that our current cultural milieu has shifted, however, from this logic of synecdoche into a realm where the parts themselves become (rather than represent) the whole. In order to illustrate, Wegenstein draws upon advertising images. In one example, she describes an advertisement depicting a nude female torso photographed in profile. Wegenstein suggests that in this particular advertisement, the breasts overcode the body in a way that the face previously had reign.[96] Instead of referring to the body of the woman, the breasts become all that woman is. She further suggests that this new era is one in which we have organs instead of bodies. That is, individual body parts have the representational power to replace the whole body without a sense of lost unity or incompleteness.[97] Parts become the whole. Unity is not referred to by the body part, but is actually constituted in the body part.

The lesson learned from Rosie, the Marlboro Man, and UmJammer Lammy confirms that the territorializing discursive practice of body part fragmentation is alive and well in body imaging. The additional insight these figures provide, however, is that this fragmentation does not necessarily mean that the arm or hand, for instance, overcode the body. Despite the beheaded nature of the Lammy advertisement, there is not quite the suggestion that Lammy is nothing but belly. This is, perhaps, partly because the tattoo interrupts that overcoding by adding another layer of signification to the image that suggests the rock star persona. Nor is this a dissected body that necessarily longs for or distrusts the totality of the body. In the cases of Rosie and the Marlboro Man, the body is not, after all, absent from the picture. It is also not a fragmentation that is to be liberated by the multiplicity of the body without organ(ization). All of these images are still quite distinctly "normal" human bodies. They are not pastiche-like creations or collages of randomly assembled parts.

The possibility for disruption and agency, rather, occurs in the points of tension between the fragmented parts and the impossibility of an easily discursively unified whole. Images of tattooed bodies are images of tattoos and/on body parts. At their agentic best these images simultaneously focus our attention on the tattoo, on the body part, and on the body as a whole without inviting us to resolve the distinctions or diffuse the tensions operating between (the gendering of) the tattoo, the body part, and the body. The tension between the gender codings implies a type of disruptive fragmentation. A fragmentation that does not act to refer outside of itself or to overcode, but one that, in MacKendrick's terms, "leaves language and body open along the lines of rupture."[98] This is a fragmentation where the lines between the parts are not mended or fused, but yet they also do not refuse the possibility of a unified whole.

Visualizations of Frankenstein can perhaps be the figure for fragmentation that leaves the body open to rupture. John Stratton, in his comparison of the monster in Mary Shelly's *Frankenstein* and the one shown in the *Rocky Horror Picture Show*, argues that the latter character lacks the visible scars and sutures that are so disturbing in the representation of the original monster.[99] The terror of the visual presence of Victor Frankenstein's monster, the source of the disruption, is that the parts of his body never fully unite into a whole because his seams are showing. The parts do not quite fit together in a fully mended whole and therein lies the power of their disturbance. The discursive force of fragmentation in controlling the body through dissection (as discussed in chapter two) is, therefore, also productive of the disruption of recombination. It is the inherent particularization of fragmentation discourse that produces the possibility of disruption and rupture because of the ongoing practice of dissection and insufficient (re)combination. The point here is not that fragments are haunted with the lack of a whole, or that they are necessarily overcoding the whole, but that the fragments *can never* be fully whole but yet they somehow are. The Marlboro Man's hand and Rosie's biceps standout of the image because of the tattoos but are also specifically not completely disconnected from the totality of the represented body. The biceps are marked as masculine and separate from the feminine Rosie head but are still somehow uneasily connected. The potential for agency within visual and discursive fragmentation is, therefore, to consider fragmentation not as "the anatomy of dissection, the dialectical dismembering of organs and functions, but rather the anatomy of configurations, of shapes."[100] The life in the image and, consequently, the agency of the images' "as-if" exists in the incipient possibility of recombination produced specifically through the territorializing effects of fragmentation and particularization. One might say that in order to have a line of rupture (flight?), there must first and continually be acts of normalization, meaning creation and reproduction, visual and discursive territorialization through fragmentation and dissection. The agentic force of the Rosie poster (and even the Tampax and UmJammer advertisements), therefore, emerges through a history of dissection and recombination, territorialization and reterritorialization—a history that has been discursively repeated throughout the genealogy of the tattooed body in representation and performance.

CHAPTER FOUR

MODERN PRIMITIVES: EXOTICISM, HYBRIDITY, AND PHOTOGRAPHY

> In the absence of truly unique, first-person experience in one's own RNA-coded memory cells, how can one feel confident about one's basic "identity"? And by extension, how can one, lacking unique experiences, create something truly eccentric? Virtually every experience possible in the world today—from touring Disneyland to trucking on photo safaris in Africa—has already been registered in the brain through *images* from a movie or TV *program*—(an apt word indeed). (We are programmed, but for what? Where does the image end and reality begin?)[1]

THIS EXPRESSION OF THE DISTRUST OF THE IMAGE as a false representation of reality that stands in for true experience is found in the introductory words of the editors of *Modern Primitives*. Originally published in 1989, *Modern Primitives* is a collection of interviews, illustrations, and photographs that presents "modern primitivism" to the viewer/reader as a revival of "primitive" (and often extreme by current Western cultural standards) body modification practices. *Modern Primitives* encourages readers "to see in their tattoos a primal instinct and even a physical force."[2] Largely due to *Modern Primitives* and other publications like it, tribal tattoo designs in the United States signify more than an artistic style. Modern primitivism, constructed through the aesthetic of specific body modification practices, became an identity performance seeking to "return" to the purity of mind and body imagined in various "native" and "primitive" cultures. In the decade following its publication,

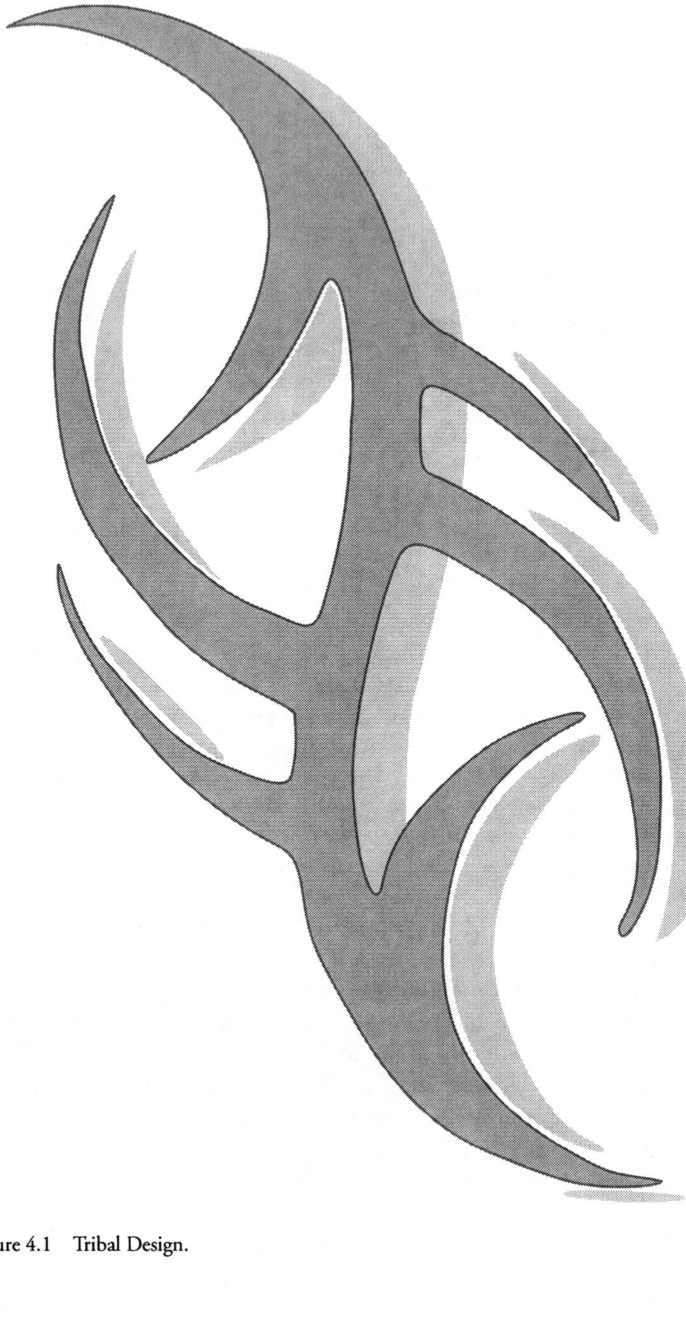

Figure 4.1 Tribal Design.

Modern Primitives, moreover, was followed by a series of books espousing this modern primitivist philosophy. *The Customized Body* (1996) and *Return of the Tribal* (1997) also articulate the modern primitivist philosophy through a combination of written text (essays and interviews), photographs, and illustrations. *Modern Primitives*, therefore, marked a pivotal transition for the meaning of tattooing in general and tribal designs in particular in the United States. A transition, it is important to note, that relied upon the force of the imagery it disavowed.

The dependence upon visual representation to express the philosophy of modern primitivism in *Modern Primitives, The Customized Body, and Return of the Tribal* stands in stark contrast to the sentiments written in the introductory pages of *Modern Primitives*. It is within this apparent contradiction between a disavowal of the "fake" image and the simultaneous dependence upon the visual, however, that the vital performativity of the exotic tattooed body emerges within the context of modern primitive visual representation. That is, the tattooed modern primitive body is constituted in and through the citation and iteration of a "real" identity created by image-text relations. That manufactured "real" identity then both authorizes and is simultaneously created by modern primitivist discourse. As we shall see, the disruptive potential of the image in this representational context, like the disruptive class and gender potential, emerges out of the tensions produced within this authorizing (and territorializing) discourse.

This chapter explores the visual representation and construction of the tattooed body in the context of illustrated books, modern primitivism, and the discursive strategies of exoticism and exotic representation. I argue that the modern primitive body produced in popular collections of modern primitive photographs is the product of the hybrid process of exoticism. I indicate, furthermore, that the historicity of late twentieth-century performances of the modern primitive exotic as constructed in these texts recalls/creates three distinct genealogical chains: the imagined history of the "tribal," the history of sideshow performances, and the marginalized working-class history of the tattooed body in the mid-twentieth century. I suggest that these histories are hybrid and, thus, partially constructed—unfinished and incomplete. It is, furthermore, because of and through the construction of this discursive and visual hybridity that productive and disruptive tensions emerge. Agency consequently acts *within* internal, not excessive or external, fissures and thresholds created in exotic territorialization.

EXOTIC PERFORMATIVITY

The production of the exotic constitutes the discursive frame for the analysis of these texts. At least since the publication of Edward Said's *Orientalism*, the issue of the appropriation of non-Western culture and cultural imagery through the process and discourse of exoticism has been thoroughly explored.[3] In particular, tattoos have been understood through this discourse since at least the late eighteenth century.[4] Despite the fairly common mobilization of this term in reference to the practice of representing cultural otherness, however, definitions of what constitutes the exotic continue to vary. On the one hand, exotics are cultural objects or a "special form" defined by "popular notions of fascination and desire: the desire to enter forbidden territory, whether in the imagination or physically, to partake of otherness and stake a claim."[5] On the other hand, exoticism is a process of isolating and making strange particular elements of a source text, object, or person in a way that focuses on the "strangeness enacted by difference that stands in as a representation for the whole."[6] The exotic, consequently, is both an object consumed as culturally different and/or dangerous, and a metonymic construction of otherness that produces that object's exotic effect. To employ a fairly literal example, consider the consumption of exotic regional or "ethnic" specialty foods. Nearly every Christmas my Scandinavian heritage is displayed on the dinner table in the form of lefse (a tortilla-like potato bread) and lutefisk (cod cured in lye and then boiled and/or baked). Eating lefse and lutefisk in my family is an act of performing as well as consuming our family tree. For guests and in-laws, the decision to eat (or not) the flat bread and gelatinous fish is wrapped up in issues about culinary exploration and "trying something different." These food items, moreover, have come to stand-in for our family's imagination of Norwegian-ness.

There is also a disagreement between cultural critics who argue that Western fascination with the exotic expresses the power of Western hegemony to colonize the cultural "other" via cultural appropriation, and the advocates of the exotic's disruptive potential. For example, some cultural critics argue that the Western enthrallment with the exotic appropriates the meaning of a culture by decontextualizing the cultural artifact or ritual.[7] This happens in classic American Western films, for instance, when Native American culture is reduced to feathered headdresses, stern facial expressions, and folded arms. The effect of such appropriation is a violent homogenization of culture casting the West as the aggressor. In Said's *Orientalism*, for example, the West relentlessly disempowers Eastern culture: "The Oriental is depicted as something one judges (as in a court of law), something one studied and depicts (as in a curriculum), something one disciplines (as in a zoological manual). The point is that in each of these cases the Oriental is *contained* and *represented* by dominating frameworks."[8] The logic that grounds the argument about appropriation is that of representational mimesis. In order for an artifact or value to be appropriated or decontextualized, in other words, there must be some specific place or pure cultural origin that is being both imitated and, through imitation, represented, disciplined, and judged.

As illustrated in chapter one, however, the logic of mimesis that supports the idea of representational reproduction or excess is limited in its ability to account for the agency of all images.[9] Moreover, this logic (and its dependence upon both the a priori exoticism of the object and of the foreign location) does not fully capture the performativity of exoticism. The point is not to deny that cultural appropriation exists (indeed, the cowboy and Indian films were made). Instead, the objective is to not limit exoticism only to this logic. Peter Mason argues in *Infelicities* (so named in reference to Austin's speech act theory), for instance, that an object is not exotic prior to its representation, nor does this object always hail from any specifically identifiable place. Rather, Mason suggests that "the exotic is the product of the work of exoticization."[10] Exoticism is, to put it another way, a "representational effect."[11] The feathered headdress is simply a garment or decoration until it becomes a sign of exotic "Indian" otherness within the frame of a film or television program. The exotic is an invention made possible through the process of performative iteration and reiteration. The production of the exotic, therefore, occurs through specific practices of representation rather than through imitative cultural appropriations. The exotic, like gender and class, is produced through its performance: it is the result of the process of exoticism.

It is not only the exoticism of the object or image in representation that is produced through this performative process. The cultural origin or location from which the object is thought to have been displaced is also a discursive production. "In this respect," Mason argues, "exoticist representation can be seen as the opposite of Edward Said's 'Orientalism'; although the latter fetishizes authoritative knowledge and geographic specificity, often with a view to actual domination, exoticist representation is indifferent to ethnographic or geographic precision and tends to serve imaginative rather than concretely political ends."[12] Objects and representational artifacts deemed exotic draw their exoticism by referring to and also creating a sometimes imaginary construct of a cultural other. As Frederick Bohrer notes, the reference culture is consequently produced through and by the cultural discourse and context from within which the exotic image emerges.[13] When I was growing up in the 1980s in a small town in northern Minnesota, for example, local tourist gift shops sold a variety of "Native American" items. These mostly included any number of decorated moccasins and faux-leather shirts and pants with extensive (often) turquoise beadwork and copious amounts of feathers. These items point to no "real" or "pure" Native American experience or culture in the late twentieth century but help to construct and perpetuate a Disneyfied cultural imagination of the American "Indian." The reference point of these object's performance of "Indian" identity is constructed through their visual display and the cultural imagination that display constructs.

What Bohrer's analysis suggests, furthermore, is that the formation of the exotic offers much more insight regarding the culture that produces it than it does about any (imagined) foreign culture. One could learn from the "Indian" gift shop much more about the culture of tourism or about class norms in my small largely working-class and service-industry-based home town than about Native American culture. Instead of focusing on the degree of fidelity between the exotic representation and that which it purportedly represents (a connection that is suspicious at best if the exotic creates what it represents), the target of analysis becomes what can be learned about the culture that produces the exotic. In order to theorize the power of the visual within this framework, moreover, yet another level of specificity is necessary. It is one thing to identify Western cultural values and norms through their representations and constructions of exoticism, it is yet another to show how visual representation operates discursively to create, reiterate, and disrupt those values and norms. This chapter consequently examines how the discursive image-text relations within the representations themselves operate to performatively produce the territorialized "foreign" spaces and objects that authorize and are authorized by Western (specifically U.S.) values and norms.

Central to that analysis is an understanding of the discursive strategies within exotic performativity. The characteristics that constitute exotics and the process of exoticism revolve around displacement (the exotic object is out of place), detachment (the exotic object has been removed from its rightful or original location), combination of sameness with difference (the object in its new place ends up verifying the values and norms of the new cultural location through its representation of cultural difference), and the superimposition of multiple meanings (as a result of the displacement, it is now an object subject to interpretation).[14] Taken as a whole, these characteristics suggest that the overall discursive structure that the exotic produces and relies upon is the object-place or part-whole relation. Sometimes real and sometimes imagined aspects of a culture become represented by objects that, once put on display in a new context, come both to stand-in for the cultural characteristic as well as eventually to perform a version of the "original" culture that is, quite bluntly, a fabrication. In short, the object (lefse and lutefisk, or the moccasins in my hometown gift shop) both operates as a synecdoche for the culture (represents "Norwegian" or "Native American") as well as metonymically performs an imagined version of the culture (Norway as Old World folk and "Indians" dancing and chanting around a fire).

In terms of the modern primitive discourse, the image of the tattooed and otherwise radically modified body is produced as a (post)modern exotic object through the process of representing and creating an imag(e)ined primitive body and the place from where tattooing originated as a technologically uncorrupted spiritual practice. The primitive object is displaced from this "original" culture and replaced in a new representational context. Moreover, as we shall soon see, the modern primitive body is out of place even within its own imagined context.

That is, these tattooed bodies belong neither to their primitive source culture nor to their contemporary cultural milieu. Additionally, the "primitive" body represented in these texts tends to be constructed as a series of illustrated, photographed, or verbally described body parts and modification practices representing the whole of exotic "tribal" identity. Different parts and combinations of parts stand in for (construct) the whole.

Exotics are, therefore, twice territorialized in order for their effect to be realized. First, the tattooed body becomes a stratified object (trace) given for representation and appropriation that has been torn asunder from two specific (although imaginatively produced) locations. Second, this trace is constructed from a series of fragmented parts and practices combined (as will be become apparent) into wholes that are incomplete and fractured. The force of the image's performativity, its agency, therefore emerges within the discursive environment that produces concrete but partial territorialized objects situated both within and between multiple discursively imagined locations/territories. The exotic image's agency, in other words, emerges from discursive hybridity.[15]

Hybridity, in the sense that I am mobilizing the term, refers to the *process* of splitting and of recombination that produces exotic hybrid representational objects. Hybridity also refers to the process of creating the imagined histories from whence these hybridly produced images hail. Analysis of cultural hybridity, therefore, is not limited to the description and interpretation of visual representations of objects that combine different elements into one unified whole. Understanding hybridity also requires attending to the process of *splitting*: a repetitious production of difference through the part-whole, object-place representation of hybrid unity. For this formulation of hybridity, I am drawing from and adapting Homi Bhabha's discussion of colonial mimicry and its production of ambivalence. To illustrate, it is worth quoting Bhabha at length:

> It is from this area of mimicry and mockery, where the reforming, civilizing mission is threatened by the displacing gaze of its disciplinary double, that my instances of colonial imitation come. What they all share is a discursive process

> by which the excess or slippage produced by the ambivalence of mimicry (almost the same, *but not quite*) does not merely "rupture" the discourse, but becomes transformed into an uncertainty which fixes the colonial subject as a "partial" presence. By "partial" I mean both "incomplete" and "virtual." It is as if the very emergence of the "colonial" is dependent for its representation upon some strategic limitation or prohibition *within* the authoritative discourse itself.[16]

While it is not my contention that the exoticist representations of tattooed bodies within modern primitive discourse are in any way comic or an act of mimicry, the ambivalence that Bhabha describes here is relevant to the disruptive agency of modern primitive representation. By trading the phrase "modern primitive" for the term "colonial" in the excerpted passage, modern primitive exoticism becomes a partial and incomplete performance whose limitations exist within (and not outside of) its own discursive production. As Bhabha notes, the transformative power of representations that partially imitate (metonymy instead of mimesis, part object-place relations instead of copies) is that there emerges *within* this territorializing process the possibility for transformation and disruption. Moreover, this disruptive and transformative potential is not merely a matter of excess or discursive rupture, but is, in fact, a latent "uncertainty" that depends upon and resides within its discursive representational process. It is within the necessarily imaginative and now incomplete performance of exotic identity within these representational contexts, therefore, that the potential for performative disruption emerges. The hybrid, in this sense, is less a border between different histories and identities than it is the difference that emerges within the territorialized constructions of identity.

In order to establish the modern primitive tattooed body as an exotic phenomenon, to discuss the resistant and recuperative performative power of its representation, and to locate the potential moments where the image of the exotic tattooed body acts out in that hybrid process, the analyses in this chapter follows three trajectories. Initially, I argue that the form and content of the three texts under consideration hybridly produce an imaginative history of primitive culture from which they gain cultural authorization. Once the hybrid representational form and process is identified, the chapter then moves on to argue that the modern primitive constructed in these texts is additionally authorized by the history of early-century sideshow performances of the illustrated body. The citation and construction of these simultaneous and occasionally incompatible histories (imagined primitive and sideshow freaks) begins to produce moments of agency. To further the analysis, these moments are illustrated in the third section that shows how the hybrid construction of multiple historical authorities, and the discursive effort to enfold one history into the other, produces the potential for particular aspects of the images to act out of the exotic/tribal context. Specifically, the analysis suggests that modern primitive discourse (because of the multiple and incomplete histories it performatively enacts) produces the working-class tattooed body as a site of disidentification because of the effort to enfold tattooing's working-class history into a territorialized narrative of neotribalism. The working-class tattooed body, subsequently, emerges as a strategic limitation of the exotic modern primitive.

IMAGINED HISTORIES: PRODUCING THE UNREPRESENTABLE

The tattooed modern primitive as an exotic body in the process of exoticism is performed within the representational format of *Modern Primitives*, *The Customized Body*, and *Return of the Tribal*. The process of exoticism enacted by these texts' representations of the modern primitive body creates the imagined history of the "tribal." Performances of modern primitives therefore gain their authority through the citation of that fabricated history. Moreover, the modern primitive is, before anything else, a body in transformation. The modern primitive body, therefore, is elusive because its source of authority does not exist prior to representation, and it exists only in its hybrid process of creation.

The following analysis of *Modern Primitives*, *The Customized Body*, and *Return of the Tribal* moves through each text individually before drawing general conclusions about the presentational form and the possibilities for agency. Throughout this analysis, I describe how these books combine written text with illustrations and photographs of Western and non-Western "tribal" bodies. The "tribal" bodies are often nude, pierced, tattooed, and/or intentionally scarred. The combination of these images with images of Western bodies and written text produces modern primitivism. The following analysis of the books as performances of modern primitivism performativity, accordingly, pays special attention to the textual rhetoric, the form of photographic representation of both the modern primitive and "tribal" body, and the histories imaginatively remembered and created in those representations.

MODERN PRIMITIVES

Fakir Musafar

Modern Primitives is a series of interviews with a variety of people who are extensively tattooed and otherwise bodily modified. The text opens with an introduction that orients the reader to the editorial view regarding the social and cultural worth and purpose of body modification. The first and longest (30 page) interview in the text is with Fakir Musafar, the self-proclaimed originator of the term "modern primitive" and "Father of the Modern Primitive Movement."[17] Musafar's body, his ideas, and perspectives are the gateway and the subsequent interpretive frame provided for readers as they navigate the world of the text. To a large extent, therefore, Musafar's representation performatively constructs the imagined "tribal" culture that authorizes much of the modern primitive discourse in the text. Throughout the interview with Musafar, the reader encounters a variety of images juxtaposing his white body with images and photographs of so-called primitive bodies. The "natives" pictured represent a variety of different cultures. They also display multiple practices of body modification.

The following description of one page of the interview illustrates how the rhetoric and composition of the text constitutes the representational effect of the exotic modern primitive body and the imagined construction of "tribal" culture. On this particular page, Musafar is asked about the basis for his tattoo designs. His response begins to create the multiplicity of the modern primitive body and of "tribal" authority:

> Inner visions. Similar, I suppose, to Balinese textile patterns that came from altered states and trips to another world. Same for the one on my back. Any tattoo that didn't come from inside you is not for you. Sometimes the artists in primitive cultures were shamans. They envisioned the marks, tattooed them on the body, and then the person who got the tattoo was whole. It was their pattern, their mark, and without that mark, they were incomplete. That's part of the magic of the tattoo.[18]

Musafar suggests that tattoos simultaneously reflect and construct tribal identity. The mark of the tattoo, as envisioned by the spiritual leader of the community, finalizes and territorializes the form of the body as an expression of individual and community identity. Tattoos make people "whole." Tattoos, far from being an imposition inscribed upon the natural body, consummate the spirit/body relationship.[19] Rhetorically constituted in this manner, adaptation of aesthetic tribal tattooing practices becomes a realization of "inner" individuality rather than an appropriation of aesthetic imagery. Tattoos are, furthermore, magical. Their ability to alter or transform the fragmented individual into a "whole" self defies explanation. This rhetoric attempts to disconnect the tattoo from its heritage as a sign of a particular social position by arguing for its ability to transcend rational logic. Instead of being either an artistic form or a signifier of social status, therefore, the tattoo is a supernatural agent. For the modern primitive, then, the tattoo artist as tribal shaman produces the body as an aesthetic expression of interior subjectivity. The modern primitive body is an expression of identity cooperatively produced (by artist/shaman and recipient) and realized in the act of production. The body is the product of both the self and the other, the expression of both internal subjectivity and external representation, both modern and primitive.[20]

The performative composition of the modern primitive is also visually enacted. On the same page that this explanation of the power and purpose of "primitive" practices of tattooing is found, appear two black and white reproductions of photographs of "native" people with tattoos. One picture displays a woman sitting cross-legged on the ground and cradling an infant in her lap. Her short-sleeved, (apparently) white, Western-style blouse appears too small for her and is completely unbuttoned, revealing her breasts. Her entire body, face, and hands are covered in a solid black-line abstract tattoo design. The caption reads, "Papua New Guinea woman with tattoos."[21] Next to this image there is a photograph of the tattooed bare thighs of a kneeling person who seems to be pulling up some kind of a garment to reveal the horizontal black-line design. The caption of this picture reads, "Eskimo Indian with tattoos made by pulling sooty thread through skin."[22] The mixture of photographic cultural references on this page and their relation to the verbal text creates the performance of modern primitive identity. The verbal caption anchors the visual image by determining its interpretation.[23] The people and body parts are evidence of ethnic others that are part of the totalized family heritage of modern primitivism.[24] The image/text becomes the stage upon which modern primitive identity is produced through interaction amongst multiple photographic images of "tribal" body parts that stand-in for and constitute character types. The visual image thus created through this metonymic process invents the reality of modern primitive identity on this photographic stage. In other words, the photography as a representational form actively produces the modern primitive. Modern primitive identity is thus exotic because it "is not originally located somewhere else and then secondarily reflected in representations. Rather, it is the product of those very representations, produced through the process of exoticism."[25]

Margo DeMello argues that in *Modern Primitives* this strategy of visual and narrative comparison between modern primitive bodies with tribal imagery succeeds in making explicit the connection between tattooing and primitive people and practices.[26] The performative effect of the composition of the text, however, actually dissolves that connection in the process of construction because what it ultimately succeeds in doing is producing an image of "tribal" culture and "primitive" practices that is not culturally specific. The form of the text creates an exotic image, a "realia drawn from regions thousands of miles from one another" and "juxtaposed in strange company with one another."[27]

The hybrid positioning of part-objects within this representational frame both constructs a modern primitive history as well as opens up a time/space that questions the authority of that history; the "tribal" body parts refuse to be completely normalized within modern primitive identity. Far from locking the body in a prison of appropriation, this hybridity "is more evocative and flexible than the specificity of a precise location in space."[28] The composition of these texts, therefore, is authorized by an image, an ephemeral surrogate, of multiple prior actual and imagined cultural performances.

This imagined community of the "tribal" is then hailed by the Western body of Musafar as his authorizing discourse in his reenactment of "primitive" practices. Musafar says, for example, that his body expresses the modern primitive desire of a "non-tribal person" to respond to "primal urges" and to do "something with the body."[29] The performative power of this image realia lies, consequently, in its location in the hybrid time/space between two worlds, one ancient and imagined and the other modern (and imagined). As Kobena Mercer's discussion of the blended African and European elements in Rotimi Fani-Kayode's photography suggests, "worldy codifications of difference are freed up by *movement between* the fixed polarities" (my emphasis).[30] The distinction between ancient and modern within the discursive and visual construction of modern primitivism is, in these terms, what produces the possibility of the movement and, consequently, gives the part-images their legs. It is through the hybrid combination and the metonymic logic within this process of constituting modern primitive identity that spaces emerge for difference to speak out.

The complex combination of multiple non-Western cultural references visually and verbally present on this one page of the interview with Musafar demonstrates the hybrid multiplicity of images superimposed upon the modern primitive body of Musafar. He is both the product and the producer of a conglomeration of cultural practices and images that, taken together, constitute the imaginary "tribal" culture that the modern primitive cites. In other words, he is the modern primitive product of the process of modern primitivism.

Leo Zulueta

The interview with Leo Zulueta is also important to *Modern Primitives*' construction of tattooing as an exoticist practice.[31] Although the interview with Zulueta is only four-pages long compared to Musafar's thirty pages, Zulueta is a tattoo artist and collector who has been a significant force in the popularization of "primitive" tattoo design.[32] His ideas and his body, moreover, are specifically dedicated to the art and practice of tattooing (rather than to body-modification practices in general). In the text, Zulueta's physical form is juxtaposed against drawings of Micronesian designs.

The expansive Micronesian tattoo design featured in Zulueta's photograph covers nearly his entire back. Upon turning the page, the reader encounters the tribal juxtaposition in the form of a graphic representation of similar Micronesian body tattoos "from an early 1900's article."[33] The illustrations are black ink drawings similar in design to Zulueta's back. The drawings are sketched on an outline of a (presumably male) human body. Zulueta explains the lack of a photographic depiction of Micronesian tattooing in the accompanying interview text:

> The last man to have a back piece like mine, who was over 90 years old, passed away a couple years ago. This is why I really feel strongly about preserving those ancient designs; besides being original art, they might be talismans for the future, or perhaps encode some cryptic knowledge that could be valuable or illuminating in some way-who knows? But if they're not preserved, we'll never know.[34]

Zulueta self-reflexively identifies the cultural memory his body kinesthetically imagines. At the same time, he increases the distance between modern primitive practices and any geographically or culturally specific group because he suggests that these practices are no longer performed in the cultures where they (presumably) originated. Zulueta's body, in its preservation of the sign of the Micronesian tattoo, also displaces that sign upon a non-Micronesian body. The image and the body are reconfigured from being either Micronesian or not-Micronesian (the polarity inherent to the logic of appropriation) into the body of the modern primitive that is both and neither. In other words, the practices of this modern primitive rehearses the disappearance of the practice in Micronesia as well as produces the disappearance of Zulueta's unmarked and non-Micronesian body.

Moreover, Zulueta suggests that the signifying potential of his body is open. His body may, for example, be the vessel for some as yet unknown source of knowledge. The point is not that the true meaning of his body will someday be discovered, but that inherent in the visual design is latent power with immanent potential for activation. Like an ancient language for which we have no dictionary, Zulueta's tattooed body may provide us with answers to questions we have yet to ask but for our inadequacy to unlock the code. Additionally, the language is not dead because it lives visually on/in Zulueta. Unlike the "natural" body whose code is easily accessed, therefore, Zulueta's tattooed body's meaning is indeterminate and immanent.

DeMello argues that Zulueta's rhetoric is an attempt to justify cultural appropriation.[35] DeMello suggests that Zulueta's physical adoption of the Micronesian images is similar to the practice of taking objects from a culture and putting them on display as artifacts in another. The source culture is, in this scenario, objectified and normalized by a (usually) Western oppressor. The difference between the process DeMello describes and Zulueta's practice is, however, that Zulueta's corporeal body is permanently altered. He transforms into the object of display. Zulueta's body produces modern primitive identity, in other words, by acting as a cultural surrogate for the Micronesian body. In this sense, Zulueta's body cannot be easily identified as a symbol of simple cultural appropriation because he is not a representation of either culture; he is (corporeally) both.

The other tattoos on Zulueta's body further contest the representation of Zulueta as the vessel of appropriation. Zulueta's arms are tattooed in "tribal" style but they are of no specific cultural origin. Zulueta explains: "I told Ed I wanted to get a big tribal-style flame on my right arm. He said, 'You draw it, bring it to me, and I'll put it on.' That was in 1978. Then I got the matching piece on my left arm in early 1981."[36] The specificity of cultural authority that, in Orientalist logic, suppresses the source culture is problematized because there is no single source that is being appropriated. The layers of meaning heaped upon Zulueta's body resist the logic of appropriation because they beg the question of what culture is victimized: is it the imagined culture created by the combination of cited "tribal" signifiers, or is it Zulueta whose body has been (permanently) claimed by the images?

Return of the Tribal

Return of the Tribal's composition is similar to *Modern Primitives*. The major distinction is that the primary written text is not interview-based but the author's effort to "show that the impulse to shape one's body and one's self in one's own desired image, far from being something only for social outcasts, seems as intrinsic to being human as are (self)consciousness, art, communication, and sexuality."[37] To that rhetorical end, the book is divided into seven chapters that combine written text and photographs of primarily Caucasian tattooed and pierced Western bodies set against photographs of "tribal" people exhibiting a variety of styles of body modification.

This style of photographic presentation begins in the "Introduction." The layout of page six, for example, shows the typed (black letters on white page) time-line of "Body Adornment and Modification" jumping from "8–5000 B.C.E." to "7000" to "4200," and so on. The time-line on this page ends up at 500 BCE, which is designated as the "Approximate date for the archeological find, in the Altai Mountains, Siberia, of the mummified skin of a male hand, tattooed with mythical animals."[38] On the opposite page, there is a colorful photo layout. The top of this page is covered by an approximately 4" × 6" close-up of the head and face of a dark-skinned "Hadendawa" woman wearing large, flat, gold hoop earrings and a gold hoop nose-ring through the top and front of her right nostril. She is wearing what appear to be two shawls over her head, the orange material of one mostly obscuring the turquoise of the second underneath.

The Hadendawa female is juxtaposed to a medium close-up shot of a white woman's head and nude torso on the lower right corner of the page.[39] Appearing in a smaller photograph than the visually older Hadendawa female, this younger woman's back is facing the viewer with her head turned and tilted so that the left profile of her face is visible and her neck is exposed. Her long wavy black hair hangs loosely over her right shoulder to suggestively reveal her bare neck and her left ear. She has two tattoo designs. The first image extends from just behind and below her left ear down her back to just below the shoulder blade. The other design runs down her left arm from her shoulder to at least her elbow (the photograph is cut off just above her elbow line). Both designs are composed of solid black sculptural lines typical of neotribal designs.[40]

Below the upper photo and to the left of the lower photo is the following text:

> Identity and Freedom
>
> A golden ring in the front of the right nostril is the trade-mark of Hadendawa women, whereas other Beja-all Muslim and heavily veiled-wear different types of nose-rings, as with other tribes across the world, people's adornments are often determined by being born into a particular tribe. Such restraints do not exist for people in non-tribal societies. The elegant and unusual tattoos befitting the elegant lady at right is a design truly unique to herself.[41]

The combination of the written text and the visual image asks the reader to make the association between the tattooed woman and the "native" Hadendawa woman.

Moreover, the caption of the written insert, like Musafar's rhetoric, draws a connection between the modern primitive performance of self and the Western valuation of autonomy. Unlike tribal culture, neotribal identity is not a genealogical imperative; it is a choice. Modern primitive and neotribal selves refer to the images of tribal practices but reconstitute them through displacement into the Western cultural context privileging individuality. The style of narrative description in the written text reinforces the displacement. When describing the Hadendawa woman, Camphausen's language is denotative and seemingly objective; she simply exists. Her life and vitality are captured and frozen in discourse as she is transformed from a living body into photographic evidence. Her neotribal counterpart, on the other hand, is "unique," "unusual," "elegant," and remains active and alive because "restraints do not exist for people in non-tribal societies." Neotribal performance creatively alters ritual markers of tribal membership because modern primitives are both uninhibited by the restrictions imposed by social structure as well as free to modify tattoo designs in order to heighten their individuality. The composition of the photographs echoes this sentiment as the Hadendawa woman is tightly controlled by the frame of the close-up shot and is encircled by not one but two shawls over her head. The modern primitive, by contrast, is free of all clothing and has more space within the photographic frame. The social affiliative element of body modification is not, however, completely recuperated by the neotribal surrogation. "Neotribal" is also an identity category and the choice to adorn oneself in that style marks membership within that group despite claims to individuality. Displaced within the juxtapositional context of *Return of the Tribal*, "tribal" body modification practices become creative assertions of individual identity as well as expressions of group affiliation.

The review of *Return of the Tribal* by *Publishers Weekly* printed on the back cover describes this visual form of the text as a juxtaposition of "photos of modern urban tribals against images of indigenous people" that "effectively places this oh-so '90s practice firmly within a historical and global context." The historical and global context, however, lacks specificity. The tattoo designs etched upon the Western body are identified as "tribal" because of their general aesthetics, not because of their reference to a specific culture or cultural ritual. Additionally, the juxtaposed images do not feature similar body modification practices. The "tribal" body is pierced and the neotribal body is tattooed. The layout and text suggest that the modern woman's body is somehow spiritually connected to the Hadendawa woman through generic aesthetics of adornment. The text also suggests a distinction, however, because the modern primitive enacts agency through individual choice and lack of social restraint. The "tribal" representative, on the other hand, is disciplined by social structure and without choice. The juxtaposition of images invites both comparison and contrast and so highlights the difference between the contemporary woman and the Hadendawan woman while drawing a connection between the two. Moreover, the text implies that the Hadendawan woman is a representative stand-in for "other tribes across the world"; she metonymically surrogates all tribal cultural performances. The Western woman then surrogates her aesthetically. The modern primitive produced by the juxtaposition, consequently, is a hybrid effect of the practice of representation and not only of cultural conflict and appropriation.

THE CUSTOMIZED BODY

The Customized Body features photographs and interviews by Housk Randall and text by Ted Polhemus. The book is separated into nine sections, each dealing with a different form of body adornment that demonstrates the human effort to "find more ingenious ways of transforming their flesh into art."[42] Each section opens with a descriptive and interpretive history of the practice under consideration before displaying black and white photographs of mostly nude individuals participating in the practice. Photographs and illustrations of "original" tribal bodies are absent from this text's performance of the exotic and erotic neotribal body. The juxtaposition between the contemporary and the "tribal" within this text is narratively, rather than visually, constructed. In the written text, tribal culture is valorized and its tattooing practices argued to be a celebration of stability and ritual. The segment dedicated to tattooing and scarification begins, for example, with a narrative discussing the origin of the word "tattoo." It provides archeological evidence of Neolithic tattooing, explains the role of tattooing for "our ancestors," denigrates mid-twentieth century interpretations of tattooing in the West, and celebrates the tattoo renaissance's "new awareness" of non-Western cultures and "new appreciation" for body decoration resulting in the choice to transform the body.[43]

Like the comparisons produced in the previous books, the narrative comparison in *The Customized Body* produces the modern primitive as a hybrid. The modern primitive body is overwritten by the discourse of tribal ritual and the discourse of choice (in addition to being physically transformed). The customized body's referent is, however, constructed completely through written narrative. The body of the "tribal" source is not visually present in the text. In its place, we are given the photographic image of the modern primitive surrogate—a modern Western body constructed by a combination of signs representing multiple cultural practices. This surrogate, furthermore, is authorized by an imagined cultural discourse.

One of these photographs pictures the volume's photographer, Housk Randall.[44] He is bare-chested and posed against a black background at a three-quarter angle to the camera. A tattoo covers the bulk of his upper right arm and shoulder. The caption reads:

> This tattoo is a visual representation of my life's path. The snake and the sun are Aztec designs—I'm half Mexican-symbolizing rebirth and wisdom through the undergoing of suffering and near-death experience. It can be read not only as a passage from darkness to light but also as a symbol of sexuality and life . . . the sperm into the ovum.[45]

Randall's discourse disperses tribal culture. The Aztec symbol of near-death experiences is refashioned as erotic and sexual. Randall produces his own meaning through this discourse. He makes his body meaningful by uttering a performative statement both citing and displacing the authority of Aztec culture. Randall's body is both different from the Aztec and, due to Randall's Mexican heritage, the same.

More than the other texts, *The Customized Body* highlights the function of photography in its performance of surrogacy. The photographer receives authorial credit and the photographs simply occupy more space in the text. The eleven-page photographic display following the two-page contextualizing essay regarding tattooing and scarification, for example, exhibits sixteen black and white photographic portraits of mostly nude male and female tattooed bodies in front of gray and black backgrounds. The modern primitive is, consequently, located in the elusive space of nowhere. Displaced from even the indeterminate visual context of photographic juxtaposition, these bodies float in a void.

Performing the In-Between

The vortex in which the bodies in all of these texts are represented, the space that configures their reality and produces their identity, is liminal space. Simultaneously this body is a visual intersection, interpretation, and adaptation of a conglomeration of "tribal" symbols that merge to form the imagination of "primitive" culture. The combination of the juxtaposition of "tribal" bodies and "tribal" illustrations with photographs of modern primitive bodies and body parts that exhibit designs that have no concrete referent, in addition to the narrative construction of "tribal" practices and societies as idyllic, produces a body that is simultaneously everywhere and nowhere. This body is neither restricted by a classified space that constrains its behavior nor does it have the comfort of a room of its own. It is, as Bhabha's terminology of the hybrid suggests, both partial as well as virtual.

The performative production of modern primitivism that these three texts create rhetorically and visually through juxtaposition of non-Western and Western bodies and body parts consequently produces a hybrid representational "third space." Bhabha suggests that the third space emerges as the process of meaning construction and interpretation that occurs between the specific act of enunciation and the "general conditions of language."[46] Bhabha notes that "the meaning of the utterance is quite literally neither one nor the other" and, as a result, there is "no way that the context can be mimetically read off from the content."[47] Within modern primitive representation, this concept is both metaphorically illustrated as well as enacted. The illustration occurs because the juxtaposition of bodies demonstrates that the exotic body is quite literally neither the native nor the modern, but some amalgamation of both. Visually, then, the representation of the modern primitive within these texts shows the reader a metaphorical version of the third space that emerges in the act of meaning construction. The texts also enact the third space because they are, of course, discursive events. In other words, the meaning of the modern primitives emerges in the third space of enunciation between the specific text and the more general discursive environment. Or, in the terms of this project, within the performative processes of iteration that occur within discursive territories. Bhabha argues that it is these "discursive conditions of enunciation that ensure that the meanings and symbols of culture have no primordial unity or fixity."[48] In other words, the relation of meaning that emerges within the performative processes of meaning construction is one that is inherently open to the possibility of transformation. Such in-between spaces therefore have the agentic potential to "provide the terrain for elaborating strategies of self-hood—singular or communal—that initiate new signs of identity, and innovate sites of collaboration, and contestation, in the act of defining the idea of society itself."[49] The modern primitive representation within the three texts considered so far demonstrates this transformative possibility by showing how the fissure of unity occurs simultaneously with its construction. The potential for the disruption of the representational context (in this case modern primitive identity) is because of (again, not despite) the efforts to normalize tribal culture into an imagined unity. It is not that the authors and editors of these volumes necessarily intended for collaboration to occur. Rather, collaboration and innovation are products of the hybrid representational and discursive strategies that break down even as they are built up. The agency of the images emerges within this unstable hybrid discursive context.

The specificity of body modification practices is also important to the hybridity of modern primitive representation. The body itself emerges as an image-text because modern primitivism is enacted through and upon the bodies of the modern primitives. The modern primitive tattooed body, as constituted by these texts, is more than a Western body overlaid with a "tribal" image. The modern primitive *becomes* through permanent physical transformation; it is literally reconfigured into a new form.[50] The possibility for hybrid transformation that Bhabha alludes to is brought into sharp metaphorical relief within the discursive and visual environment of modern primitive tattooing. Museum exhibitions may present a cultural object taken from one location and displace it into another, thus placing the object as a representation of the cultural "other." The modern primitive body, on the other hand, produces itself *as* a cultural other. It produces itself as the "in-between" space, thus eluding the politics of polarity and emerging as the other of itself.[51] The modern primitive body, therefore, functions to demonstrate both how the juxtapositional and metonymic representational process of modern primitive exoticism takes place and how the power of hybridity transforms the "body" of territorializing discourse.

Revising the Primitive: Citing Sideshow Authority

The story of the modern primitive constructed by these illustrated texts does not end with an imagined history, however. These texts both perform a surrogation of the imagined cultural performance of the "tribal" other and resurrect the performance of the tattooed exhibit early in the twentieth century. Sideshow history both authorizes the modern primitive's exotic performance as well as contests the performative authority of modern tribal imagination.

Return of the Tribal refers to the practice of displaying illustrated people in its historical time-line of body modification and adornment practices.[52] *The Customized Body* refers to this history in its mention of the association of tattooing with "freakish" behavior.[53] *Modern Primitives*, however, more clearly evokes the history of the sideshow with its representation of "Tattoo Mike" and "Captain Don." It is on those two figures, therefore, that I concentrate.

Tattoo Mike

"Tattoo" Mike Wilson, according to Vale and Juno's biographical sketch, began getting tattooed "at an early age."[54] Also according to the biographical sketch, Wilson was featured in Dick Zigun's "Coney Island Side Shows by the Seashore" as "The Illustrated Pain-Proof Man."[55] Unlike many other interviews published in Vale and Juno's volume, the three-page section devoted to Tattoo Mike is devoid of any visual references to "tribal" art or practices. Instead, there is a photograph of Wilson at the Coney Island sideshow standing bare-chested on a stage and leaning slightly over as he props up a bed of nails so that (what appears to be) an audience member can inspect it.[56]

The other photos of Wilson reveal torso views of him in front of gray backgrounds and are, therefore, similar in form to the other portraits in the volumes. However, the photograph on the first page of the section, like the photograph of the sideshow, violates the norm of representing the modern primitive bodies in void space unmarked by time. In this image, Tattoo Mike is facing the camera and lying on his left side in tall grass propped up on his elbow. He is bare-chested and is wearing a pair of worn blue jeans. Tattoo Mike's right forearm is resting lightly on his waist and he is holding a lit cigarette in his right hand. Tattoo Mike's entire bare chest, hands, arms, and face are covered in tattoos. The shot is a medium close-up with the grass forming the background and the out-of-focus horizon only slightly visible above the grass. This field could be virtually anywhere. Temporal location is more distinct, however, marked by denim jeans and a cigarette that are suggestive of the mid- to late twentieth century. The jeans and the lit cigarette (as opposed to a nude body, a body in tribal garb, or a man dressed in a suit) materialize Wilson as both exotic due to the extensive facial tattoos, and as working class and masculine.

Wilson's original inspirations for his tattoo designs, far from being tribal, were located in the circus and art worlds:

> Seeing more and more circus photos inspired me. Also, when I was thirteen I was studying a lot of Surrealist art. I saw photos of a stage production of Jean Cocteau starring a heavily tattooed man and this became a *key*, signifying a possible way of going through the looking glass for me to achieve a whole other frame of reference, and to elicit experiences beyond the "normal" . . . presenting yourself as a signal beacon drawing things to happen to you. In other words: *tattoo as a passage to another life*. However, I've never had an absolute philosophical or religious program behind what I was attempting to do. But getting heavily tattooed definitely made interesting things happen to me![57]

I think it is important to mention here that the italics are the interviewers' addition to the text.[58] The highlighted text focuses on the modern primitive-oriented discourse of tattooing as a metaphysical experience, even though I think Wilson was talking about something very different than a spiritual transformation. He indicates later in the interview, for example, that being tattooed made him a marked man in the sense of social display:

> I was definitely a marked man, and most of the time I liked it, but it does get in the way, such as when I'm looking for work. I like the tattoos themselves—I thought about them carefully and chose the tattoos I wanted. I like meeting people with tattoos—it's a weird bridge to meeting people; sometimes I'll meet people in kind of an odd way. But what I don't like is the *obvious*. I'm in theater 24 hours a day when I'm in public.[59]

When Wilson suggests that tattooing is transformative, it appears to me that he is referring to a visual physical transformation and not a spiritual awakening. Once again, however, the editors italicize the word that might cut against this non-neotribal message. That is, they highlight Wilson's disdain for the theatricality of his body on display before an audience—an audience that Vale and Juno characterize in the "Introduction" as media programmed, deindividualized, and whose minds have been colonized by TV images.[60]

Vale and Juno thus elevate modern primitive identity above the rest of society and separate its practices from conformist, disembodied, simulated, and image-driven culture.

Vale and Juno's focus throughout the publication is on the practices of body modification as "uniquely personal experiences" that "underscore the realization that death itself, the Grim Reaper, must be stared straight in the face, unflinchingly, as part of the continuous struggle to free ourselves from our complexes, to work out unaccountable aggression and satisfy devious urges."[61] Wilson's very practical outlook on tattooing challenges this ethic of the modern primitive. His perspective is perhaps most aptly illustrated in his answer to the question "What does tattooing mean to you now?" His response: "To be honest, I don't even think about it any longer."[62]

Despite the distinctly nonmodern primitivist attitudes that Wilson articulates, the editors include him in their volume. The rationale for this decision is perhaps located not in Wilson's behavior, but in his body as a visual representative of tribal art. Wilson's tattoos are not only inspired by the circus, it seems, but are also derived from Samoan and Indian designs.[63] He characterizes his body art as a "kind of *psychedelia* of different cultures combined" and says that in the planning stages of a tattoo he gets "deeply involved with the spiritual and metaphorical implications of the prospective design, but after it's on I forget about it."[64] Wilson's body is a temple of primitive images and he has just enough intellectual involvement with the designs, if not a spiritual involvement with the practice, to warrant a presence in the text.

CAPTAIN DON

Tattoo Mike is not alone in his citation of the circus sideshow history of tattooing, Captain Don Leslie is swallowing swords right alongside him. Captain Don is heralded in the text as "a survivor of the forgotten era of the traveling circus sideshow."[65] Captain Don's photographs show him to be heavily tattooed on his chest, back, and arms.[66] In the first photograph, Don is bare-chested in the act of sword swallowing. The pictures spread throughout the rest of his chapter show him onstage with the Cristiana Brothers Circus, posing with his back to the camera while holding a sword so its handle is visible over his shoulder and playing a guitar behind a three-sided circus display board announcing "Capt. Don Presents Songs & Swords. Unique Music. Circus Acts. Strange, Odd & Unusual Feats."[67] Another photograph included depicts Captain Don's sword swallowing mentors, Carlos Leal and Estelline Pike, "the last lady sword-swallower, now retired in Florida."[68]

The interview with Captain Don is dedicated primarily to his circus acts and to a type of circus and carnival history lesson.[69] Captain Don is presented as a soon-to-be-extinct icon of days gone by who knows what it was like back when tattooed people were exhibited regularly on the sideshow circuit. The introduction to his interview, for example, indicates that he is "a survivor of the forgotten era of the traveling circus sideshow."[70] As a "survivor," Don is cast as both an expert on sideshow performance as well as a fetishized relic of a gladly departed past. Vale and Juno, for example, ask a series of questions that both establish Captain Don's authority as a sideshow performer as well as suggest that he represents a (thankfully) dying or dead tradition. For instance, Vale asks Captain Don how long he has been a sword swallower, requests that Captain Don describe his act, wonders how long it took to train for the profession, and murmurs a ponderous "Hmm" when Captain Don reveals that none of his children are carrying on his tradition. Captain Don's answers to these questions form a narrative describing life as a sideshow performer, the culture of the circus and carnival, and the experience of sword-swallowing and fire-eating.

Throughout the interview, Captain Don makes no pretense to adhering to a modern primitivist philosophy. Like Tattoo Mike, he is very clear about his nonspiritual association with tattooing. He indicates that his motivation for getting tattooed was material: as a sword-swallower, human pincushion, and fire-eater, he could simply make more money with the addition of tattoos.[71] Captain Don's story is that of the decline of the power of the illustrated person to command an audience.[72] When tattooed people became commonplace, they had to learn to perform other acts on the sideshow platform. Captain Don merely recognized that being tattooed was an asset to his current act and so he "got tattooed to be on exhibition."[73] Moreover, many of Captain Don's designs are decidedly not tribal. His back piece, for example, is a Christian image: a large cross scrawled with "Rock of Ages." Nor does he position himself as an executor of primitive practices. He insists that he is not a tattoo artist but a tattooist or, in other words, "an applicator" of tattoos.[74] When asked when he met Ed Hardy, tattoo artist and author who edited the "New Tribalism" edition of *Tattootime* and is also a subject of the volume, Captain Don honestly replies that he "never even heard of him 'til '81; I wasn't around the tattooing scene *per se*; I was around the circus sideshow."[75]

The one tilt of the hat that Captain Don pays to the spirituality of tattooing is very late in the interview when he is responding to the inquiry about the meaning of early sailor tattoos. In his answer, he refers to the human "primitive instinct" to decorate "him or herself" and to body practices in Africa and China.[76] He also suggests that a tattoo can be a message from the soul.[77] On the other hand, in the same answer, Captain Don also declares that most body decoration is motivated by self-expression: "I want you to know who I am."[78] He furthermore insinuates that this desire is founded in social disaffiliation: "especially we who like to separate ourselves from the populace and be in special groups . . . "[79] These statements are much more indicative of a mid-twentieth century mentality of subcultural group affiliation than they are of the modern primitive philosophy. The ellipse, like the italics in Wilson's interview, seems to be an editorial choice. Perhaps Captain Don was coming too close to suggesting that group affiliation, as opposed to the individual identity marketed in the introduction, motivates people to become tattooed.

Part of *Modern Primitives*' anomalous performance of modern primitivism includes the performances of tattooed bodies that are a direct link to the sideshow past. Once that past is heralded, the implications of performances of tattooed freaks and their conventions of display are cited as an authorizing discourse for the cultural performance of the modern primitive. In other words, the acting bodies of Tattoo Mike and Captain Don reenact cultural performances of the tattooed body of the past and displace them not only in the context of modern manifestations of the sideshow, but also within the discourse and performance of modern primitivism.

Exotic Presentation and the Illustrated Person

Illustrated people on sideshow stages in the late nineteenth and early twentieth century performed a version of primitivism that modern primitivism displaces. Tattooed exhibits were presented as exotics designed to appeal to "people's interest in the culturally strange, the primitive, the bestial, the exotic."[80] Tattooed bodies were particularly compelling in this mode because "early anthropologists saw the practice of tattooing as the ultimate sign of primitiveness, revealing a lack of sensitivity to pain and unabashed paganism."[81] Unlike other exotics on the freak stage who were often ethnic minorities performing "wild men" and "savages," however, the tattooed exotic was usually a white man or woman who claimed to be a victim of torture at the hands of "primitive" people. The narratives surrounding the performances of these tattooed exhibits were consequently fabricated tales of kidnapping and rescue after enduring and surviving the intense pain of tattooing.

Adventure narratives of torture are thought to have originated with the introduction of John Rutherford to English popular culture in 1828: "Rutherford declared that he had been captured by the Maoris who held him prisoner for six years; he had been compelled to marry the chief's thirteen year old daughter, he had three children, and was forcibly tattooed. His story aroused sympathy and publicity, on which he capitalized by becoming a successful exhibit."[82] Rutherford's story was eventually revealed to be fiction. Rutherford was actually a mariner who had jumped ship and chosen to marry a native woman and become tattooed.[83] Those who followed him onto the stage, however, capitalized on his narrative's success in accessing the popular cultural curiosity about primitive people. Subsequent illustrated men in the United States told stories that were remarkably similar to Rutherford's. Captain Costentenus, for example, borrowed the major motifs of Rutherford's chronicles. The following account of Costentenus' trials is courtesy of his fellow circus mate, Robert Sherwood:

> Costentenus was a native of Albania, in Turkey, although his parents were Greeks of Christian faith. On one of the wild raids of Ali Tebelen, the bloody Pasha of Yanina, who attempted to press the young Greeks into his army, young Costentenus was captured . . . Costentenus fell a victim of the tattooing-needle through his Christian faith. Surely the Christian martyrs in the Roman arenas suffered no more than he. The pain must have been terrible, although not affecting the brain. He told me that as the lines of blood-drops extended farther and farther over the white sensitive skin, and all the nerves began to twitch, the sweat stood out on his brow and he groaned aloud with pain . . . After his grueling experience, his body, terribly swollen and inflamed, was wrapped in bandages, previously soaked in linseed oil. . . He was discovered, rescued and brought to America by P.T. Barnum, who learned of him through a Spanish sailor while traveling in the Mediterranean on Tom Thumb's first European tour. He was first placed on exhibition in this country at George Bunnell's Museum.[84]

Costentenus's story is remarkably similar in theme to Rutherford's.[85] Costentenus' story was, of course, also a fabrication. It turns out that he had himself tattooed with the intention of going into show business.[86] Illustrated women also jumped onto the antiprimitive bandwagon when they stepped onto the sideshow platform in the late 1800s. Nora Hildebrandt, for example, claimed that her father tattooed her under threat of death from Sitting Bull.[87]

These stories were, for the most part, fictional and acted to create and perpetuate the imagination of primitive culture as savage and uncivilized. For the popular audience, these performances and narratives established primitive practices firmly in the category of the cultural and social other. The effort of the modern primitive narrative constructed in contemporary illustrated texts, of course, rehearses the themes circulating in the early century by hailing primitive and tribal culture and adornment practices as inherently human and spiritual. Modern primitives celebrate tattooing for the same reasons that it was abhorred early in the century. Cesare Lombroso argued in 1896, for example, that "tattooing is, in fact, one of the essential characteristics of primitive man, and of men who still live in the savage state."[88] Lombroso, consequently, represented tattooing as an "atavistic revision, evidence of an individual's regression to a more aggressive, antisocial form of being."[89] Modern primitives, on the other hand, agree that tattooing is an essential characteristic of primitive culture but argue that the regression to primitivism is a positive method of reconnecting to the body through what is, according to their rhetoric, the most human of behaviors. The historicizing narrative surrounding both sets of performances is remarkably similar. The primary difference is that in the modern primitive accounts, the valued position shifts from civilization to primitive culture. It is not unusual, therefore, that the physical representations of the modern primitive would also cite the authority of the early freak displays.

According to Bogdan, some of the major conventions of sideshow exotic display included costuming and photographic representation designed to magnify and exaggerate cultural difference as erotic, exotic, and strange.[90] The costuming, for example, would be sparse and "compatible with the story."[91] Tattooed bodies were unique in some aspects because of the need to reveal the body in order to reveal the tattoos. The costume for illustrated people was, for the most part, their decorated skin. Illustrated women broke feminine conventions of physical exhibition by showing body parts strictly coded as private. Tattooed women exposed their entire legs and even appeared in public in costumes permitting observers to "look upon the quaintly decorated skin of the upper portions of the chest and back, the arms and the exposed portions of the lower limbs."[92] The female tattooed body was erotic as well as exotic, and physical exposure was critical to that performance.

The male tattooed body also performed the striptease aspect of exotic/erotic performance. Sherwood's description of Costentenus, for example, suggests that the illustrated man had to undress in front of the audience in order to reveal the tattoos.[93] The male body's exoticism was contextualized a bit differently, however. Because the norms for physical display were drastically different for men and women, male tattooed exotics were eroticized by a connection to sexual ability. The narratives, for example, reinforced the dominion of man over woman by asserting that the tortured men were given women of the "tribe" as their wives. Costentenus, moreover, was described by the sideshow talker as a "wild tattooed man" who "is much adored by all ladies."[94] In addition to the obvious gender implications attesting to masculine sexual prowess and feminine submission, this characterization constituted the primitive tattooed man as an erotic as well as exotic figure.

The pictorial displays in the modern texts' performance of modern primitivism clearly reflect this erotic/exotic connection via physical exposure. The bodies in *The Customized Body*, for example, are nude for the most part. A photograph in the tattooing and scarification section explicitly shows a man holding his tattooed and pierced penis in his hand.[95] The women photographed, on the other hand, tend to be pictured in poses that both reveal and conceal their exotic bodies. The only full frontal nude shot of a woman in this section has her covering her genitals with her hands.[96] The other photographs of the women show them in various levels of profile poses partially both obscuring and revealing their bodies. The cultural tension between exposure and concealment that affected tattooed women in the late nineteenth and early twentieth century continues to constrain the female modern primitive body, while the male primitive continues to be associated with active realization of exotic/erotic potential.

Photographic representations of illustrated men additionally authorize the discourse and performance of the contemporary modern primitive. Bogdan indicates that photographic representation displayed the freak body "in front of a backdrop depicting jungle scenes or exotic lands."[97] For example, a *Harper's Weekly* advertisement shows Captain Costentenus tied to stakes in the ground while a "native" woman tattoos him before male onlookers.[98] The images of Fakir Musafar physically remember this image. In the opening photograph in *Modern Primitives*, Musafar is hanging by hooks inserted through the skin of his chest. His head is thrown back, he has a stick (or bone shard) pierced through his nasal septum, his upper arms are wrapped in bands, and he is holding his right wrist with his left hand at his waist. The head of his penis is tied tightly with a piece of string and his hips and pelvis are decorated with a solid black abstract tattoo design following the contour of his body and appearing to extend around to his lower back. The picture does not disclose from what he is hanging so it is as if he is suspended in midair. The setting for the photo is a scene of the open plains. Musafar's body, like Costentenus's, is represented in an uncivilized or naturally primitive setting.

The physically torturous element of primitive practice enacted upon a nontribal body is, however, displaced by the rhetoric of choice. Modern primitives enact the physical rituals of pain and body modification but they are chosen behaviors, not impositions. The rearticulation of the "primitive" practices in the context of the late twentieth century consequently reflects the cultural context of the contemporary era as much as the illustrated men and women's stories participated in their cultural context. The New Age spirituality popularized since the 1960s encouraged an ethic of consciousness transformation through the adaptation of non-Western religions and practices.[99] New Age philosophy, therefore, provides part of the cultural context authorizing the rearticulation of primitive practices previously viewed as savage and bestial. Within the New Age and modern primitivist framework, pain is transcendent: "Today, something as basic as sex itself is inextricably intertwined with a flood of alien images and cues implanted from media programming and advertising. But one thing remains fairly certain: pain is a uniquely personal experience; it remains loaded with tangible shock value."[100] For the modern primitive, the pain and torture purportedly experienced by the illustrated men and women become voluntary acts of spirituality. Modern primitivist bodies such as Musafar's are consequently constituted through a rehearsal of past performances displaced within this new context. Without Lombroso's rhetoric and Captain Costentenus's performance, Musafar's body and New Ageism would lack performative force.

The performance of the body of illustrated men and women in the late 1800s and early 1900s both referred to and created an imagined version of primitive life that was tied to painful practices of body modification that were interpreted as savage and bestial. The mode of presentation of these bodies displayed them as exotic and erotic because of the narrative representation of their experiences as well as their costumes and photographic and illustrated representations. These performances were produced within a cultural context that encouraged negative evaluation of primitive and tribal cultures. The performance of the modern primitive body constituted through the texts of *Modern Primitives, The Customized Body*, and *Return of the Tribal* occurs within a new context, where the exotic tattooed body gains new meaning. In the late twentieth century, the illustrated body reappeared within the cultural context of New Age rhetoric. The early century's negative interpretation is, therefore, both cited and, through displacement in the context of New Age rhetoric, partially revised. The modern primitive body additionally refers to the performances of the illustrated men and women through their physical representations in these contemporary texts. The modern primitive is constituted as exotic partially because it visually refers to the exotic presentations of the past through costume and photographic representation, and partially because it performs a new imagination of tribal culture.

DISIDENTIFICATION OF THE WORKING-CLASS BODY

Throughout the process of recontextualizing, the performance of modern primitivism also attempts to transform the history of the working-class tattooed body through a process of incorporation. For the most part, the representations participating in the performance of the modern primitive or neotribal body enfold much of the mid-twentieth century history and interpretation of the practice and image of tattooing and tattoos into a narrative of primitivism or tribalism. In the historical time-line of body modification practices from 60,000 BCE to the 1990s presented in *Return of the Tribal*, for example, the mid-twentieth-century working-class history of tattooing is reconstructed as being the site for the enactment of subcultural urban "tribal" identity.[101] The 1950s, a decade when the image of the tattooed working-class male circulated in the Marlboro Man campaign and when tattooing was evaluated as a practice primarily associated with lower-class bodies, is described in the following way: "Subcultures such as the early rockers and teddy boys helped bring the tattoo to a new popularity. Also, the Mohawk hairstyle, based on the practice of a Native American Tribe, appeared among members of New York City street gangs."[102] The performance of seamen, soldiers, and juvenile delinquents, as well as the homosexually and culturally coded lower-class deviance that Steward and Sanders report in their ethnographic studies, have no place in Camphausen's narrative. Instead, Camphausen features punk subculture, Fakir Musafar, and the publication of *Modern Primitives*. Moreover, Camphausen highlights the Lollopalooza musical tour (begun in the 1990s by musician Perry Farrell) because, as he puts it, it enhanced the tribal renaissance through its "tribal-like gathering combining entertainment with political and human rights concerns."[103]

Unlike *Return of the Tribal*'s almost total disavowal of working-class history, *The Customized Body* does briefly acknowledge this element of tattooing's past:

> The ever-growing ranks of the Western middle-classes, however, continued to show disdain for tattooing-vehemently labeling it as "barbaric" and, worst of all, "common." Ghettoized and stigmatized in this way, tattooing in the West became associated with the disreputable, the criminal and the freakish. Certainly in such conditions the odds were stacked against it developing as an art form. Yet despite this the 1960s saw the beginnings of a "Tattoo Renaissance" which (reflecting a new awareness of and respect for non-Western cultures in general) has forced a new appreciation in the West of the aesthetic possibilities of this form of body decoration.[104]

This account of tattooing's history summarizes thirty years of tattooing in a paragraph implying that this disdainful and, by implication, insufficient representation of the tattooed body is something to overcome. Polhemus also disregards the aesthetics of tattooing in the working-class American tradition: "Western tattooing had degenerated into a kind of haphazard graffiti with hackneyed hearts, sailing ships and bluebirds placed willy-nilly all over the body."[105] This rhetoric suggests that the ritual and aesthetic possibilities inherent to tattooing overcame the disrespectful history of disdainful and degenerate art. Considering that the modern primitivist position also vaunts the permanence of tattooing, this disappearance is unlikely. The modern primitive performance cannot entirely diffuse the physical reality of the American tradition of tattooing within this territorializing narrative, however. The tattooed body permanently and visibly retains the memory of its various pasts despite reterritorializations. Despite this fact, neither *Return of the Tribal* nor *The Customized Body* afford any visual space to the histories that their narrative accounts attempt to reconfigure.

The text that arguably inaugurated the textual performance of modern primitivism and, according to DeMello, marks the movement of modern primitivism from the margin and into the consciousness of mainstream culture, on the other hand, does include visualizations of the working-class aesthetic.[106] *Modern Primitives* does not deny or deride this visual history of mid-twentieth century American tattooing, Instead, it attempts to enfold working-class history into a narrative of new tribalism and, in so doing, tries to transform those bodies into modern primitive bodies. In addition to being the heralds of the sideshow past, for example, Tattoo Mike and Captain Don also stand out discursively (despite Vale and Juno's best efforts) from the crowd of modern primitives as being of a different class identity. In their interview of Captain Don, for instance, Vale and Juno attempt to elicit an explanation of sailor tattooing that is consistent with spiritual transformation and silences the discourse of group affiliation. Moreover, Ed Hardy, who is also heralded as an important force for transforming tattooing in *The Customized Body*, is presented as a critic of popular culture and a master of Japanese tattooing even though Hardy's body is a testament to the American history of tattooing. While discussing the growth in the aesthetic potential of tattooing, for example, Hardy says, "I wish I was wearing the kind of work I can do *now*—but you only get *one* skin! And that sorta makes me feel bad, because when I'm surfing in Hawaii and when people see my tattoos, I always have to go, 'Well, this isn't really what can be done.'"[107] Hardy also makes it a point to remind Vale and Juno that he has his "credentials as an old-school tattooer—I'm part of that tradition" and even refers to some of the recent work he has done derisively as "chi chi nouveau."[108] The effort to collapse old school working-class tattooing into modern primitive territory as an artistic transformative process of self-expression that should be spiritually meaningful, as opposed to a group affiliative process occurring on a whim, cannot, however, contain the visual power of Mike and Don. In fact, it is *because* of the process of enfolding, that it becomes possible for Don and Mike to emerge as different.

CONCLUSION

Throughout this chapter, I have argued that the texts *Modern Primitives, Return of the Tribal,* and *The Customized Body* performatively construct through the process of exoticism the performance of modern primitivism. Modern primitivism thus contrived performs the unrepresentable third or hybrid space of the in-between. It is within this space and through that construction, furthermore, that the possibility for images to act through and out of this discursive and visual context is produced. That being said, it is still worth pausing a moment to acknowledge that the immanent potential for agency within the discursive territory of modern primitive representation is not always going to be realized. In some situations, for instance, the act of modern primitive representation is merely another oppressive and appropriative case of aesthetic modernist primitivism— that is, the artistic practice of appropriating artifacts and artistic practices from primitive cultures and putting them on display in modern museums as exemplars of "tribal" or "primitive" culture. It is this practice of appropriating the other and fetishizing difference to which Said objects. The "primitive" is either a source of spiritual awareness or a spectacle of savagery; in both cases, the source cultures no longer have a claim to themselves because they have been swallowed up by the West and spit out as simulation. This trend also extended into theater as tribal rituals and performance practices became the inspiration for many avant-garde modes of performance. Christopher Innes's exploration of avant-garde theater from 1892 to 1992, for example, exposes the appropriation of ritual practices as "deeply questionable" and occasionally demonstrating a "tendency towards arcane or facile exoticism."[109]

It may also occur to readers that the form of the texts in question also resembles the aesthetic of pastiche. Jameson defines pastiche as the "random cannibalization of all the styles of the past, the play of random stylistic allusion, and in general what Henri Lefebvre has called the increasing primacy of the 'neo.'"[110] The world of the "neo" is, according to Jameson, transformed "into sheer images of itself" and replete with "pseudo-events."[111] Jameson argues that pastiche, consequently, although it shares some of the formal characteristics of parody, has lost parody's "vocation":

> Pastiche is, like parody, the imitation of a particular mask, speech in a dead language: but it is a neutral practice of such mimicry, without any of parody's ulterior motives, amputated of the satiric impulse, devoid of laughter and of any conviction that alongside the abnormal tongue you have momentarily borrowed, some healthy linguistic normality still exists.[112]

For Jameson, pastiche is impotent because imitation, rather than critique, is the end result.

Neotribalism clearly falls within this definition. It is a product of the combination of multiple styles of body modification cobbled together from multiple cultures in order to create a new identity. From this point of view, neotribal or modern primitive practices become yet another example of colonizing the other via the nostalgic image culture of the postmodern. The primitive is "thus constructive, not disruptive, of the binary ratio of the west; fixed as a structural opposite or a dialectical other to be incorporated, it assists in the establishment of a western identity, center, norm and name."[113] In this sense, modern primitives are aligned with the artistic tradition of modernist primitivism as well as the recuperative practice of postmodern pastiche.

Thus conceived, exoticism is not the productive concept that I described at the beginning of this chapter, but a reduction of difference. Objections to modernist primitivism and pastiche, however, represent a narrow definition of exoticism—one that this study has rejected as a means to evaluate the specific representation of modern primitivism within the texts under consideration. My analysis of the tattooed body produced in the process of exoticism offers another option. This analysis suggests that the tattooed body as a sign of modern primitive identity, due to its specific form and context, is not a symbol of simple cultural appropriation. Instead, the tattooed modern primitive expresses the performative process of exoticism partly by representing the productive "third space."

Moreover, I contend that images can speak out through normalizing discourse. This possibility is most obviously illustrated in *Modern Primitives*. Vale and Juno's apparent desire in this text is to propose modern primitivism as an answer to a postmodern identity crisis. They define tattoos and tattooing as a means to achieve some sort of essential primal and spiritual identity. The rhetorical strategy of the authors is, therefore, a nostalgic attempt to appropriate "native" cultural practices by offering up tattooing as a technique that Western culture can use to save itself. As a result of this rhetorical goal, however, some of the tattooed bodies represented in and on the pages of this volume act out of their context because the meanings they signify are not discursively compatible with Vale and Juno's rhetoric of modern primitivism. The bodies of Tattoo Mike and Captain Don, for example, refuse to be normalized within this discourse.

This possibility for agency is produced specifically because Vale and Juno attempt to classify both the tattoo and the process of tattooing. In other words, when modern primitive discourse normalizes the tattooed body through a part-object economy of representation, the partial and incomplete discursive territory produced through this process of historical and textual hybridization creates a space for the image to speak out. The incompleteness and incompatibility of one historical explanation with the other exposes the ineffectiveness of all of these discourses to contain the meaning of the tattooed body. As parts of the whole of modern primitive representation, Tattoo Mike and Captain Don were given a crease-space to voice alternative histories. As a result, the images of Tattoo Mike and Captain Don, like the body of the unruly tattoo contest entrant, were provided with agency.

THE FOLD

The production of the fold within the hybrid discursive environment, therefore, becomes another aspect of the agency (in addition to the processes of metonymic juxtaposition and transformation mentioned earlier) produced within the image-text relations in these modern primitive texts. It is the line produced, in other words, in the relation between the narrative of modern primitivism and the visuality of the bodies of Mike and Don that destabilizes the force of the narrative to completely territorialize. I am drawing from Deleuze's discussion of the fold in *The Fold* and in *Foucault* for this explanation of the relation between the image and text.[114] Deleuze describes the relation between image and text as being between monadic, self-sufficient (territorialized) entities. These monads are physical/material and/or discursive spaces that have relatively closed boundaries. Conley explains that in *The Fold*, monadic space is represented by the artistic and symbolic function of an emblem:

> In a limited sense, the Renaissance emblem (whose principle here has been used to explicate the second chapter of *Mille plateaux*) sought to impose moral truths through the mechanism of a motto, a text, and an image; yet the relation of difference established between the visual and lexical registers effectively exceeded the emblem's objective of control, subverting it's moral or edifying aims.[115]

Within the discursive and visual structure of the emblem, the encapsulated territories of text, motto, and image collide. At those points of collision emerge points or folds of disruption.

While I am not suggesting that the images in the neotribal books function as emblems, this articulation of the relations between monadic textual and visual registers is illustrative of the image-text relations in the modern primitive books. That is, the strength of the boundaries of difference between images of, say, Don and the tribal "native" is what produces the possibility of disruption. The force of the bounded textual territory of the neotribal book to control and define what modern primitivism means actually breaks down in the boundary collision between the monadic image and text territories within that larger (book) unity. It is the sheer force of the separateness operating within the bounded territory that produces agency. Deleuze's discussion in *Foucault* of what happens when visual and language monads meet illustrates that

> Neither language nor light will be examined in the areas that relates them to one another (designation, signification, the signifying process of language; a physical environment, a tangible or intelligible world) but rather in the irreducible dimension that gives both of them as separate and self-sufficient entities: "there is" light, and "there is" language. All intentionality collapses in the gap that opens between these two monads, or in the "non-relation" between seeing and speaking.[116]

The efforts of Vale and Juno et al. to enfold the visual presence of, for example, Mike and Don into the territorializing narrative of modern primitive identity creates, in these terms, a fold because of the incommensurability of the two monadic territories: the image and the discourse.

Deleuze's argument, furthermore, is that through this understanding of the fold "intentionality collapses" in such a way that "casts off" psychologism.[117] In this case, the intentional effort of the authors to create a territory within which Don and Mike are normalized through the interpretation of their tattoos as the evidence of some spiritual/psychological essence of human identity is overcome by the force of the monadic collisions *within* the textual and discursive formation. Part of the process of hybridization, then, includes the production of disruption within the lines of contact that emerge from the collision brought about through the folding of the visual within the discursive. Moreover, the efforts of these texts to create modern primitivism partially through the erasure of the popular cultural past of tattooing in the United States cannot deny the visuality of that history written upon bodies. By discursively identifying itself in opposition to the working-class tattooed body, the modern primitive places the working-class body in the position to destabilize modern primitive discourse. In other words, the working-class body wearing an assemblage of old-school tattoos is provided with a gap-space to talk back to the modern primitive because of the act of discursive enfolding and not in resistance to or despite that act.[118]

Modern Primitives is not, therefore, a neutral text. The exotic production of the modern primitive tattooed body both fails to fully achieve the authors' goal of recoding tattooing and, because of that failure, reveals the agentic potential of the image within this discursive image-text environment. Incompatibilities erupt on the pages that beg the question of what is being "cannabalized": is it "native" culture, working-class masculinity, freak shows, or the modern primitive's body? In the process of containing and subsequently enfolding all of these individual territories of identity, history, and discourse together within the boundaries of the larger territory of the bounded volume of a book, folds and creases of collision are created where agency emerges.

Conclusion: Imag(e)Ining Agency and Criticism

This project began with the desire to discover moments of agency within the performative production of meaning in and through visual images of tattooed bodies. The purpose throughout has been to perceive occasions when the normalizing force of meaning fails to erase particularity or difference because of (and not despite or in resistance to) its performative (re)production. What I have proposed and subsequently sought to demonstrate, therefore, is that the agency of the visual works, at least in part, through the power of the image to performatively act out of and through stratified structures of discursive containment and finalization. In each of the contexts of visual display featured in this text, discursive constraints produced the possibility for images to move and/or to rematerialize within stratified discursive territories. In other words, one of the ways an image's "legs" can get moving is through and by the performative production of meaning.

The figures for these moments of agency in each of the contexts echo, furthermore, the spatiotemporal conditions of the assemblage as a network of moving relations operating on multiple planes at differential speeds. As suggested in chapter one, there are various qualities of relations between and of time and space. Occasionally events slow down to the point at which time and space are frozen or appear as a static and immutable location. It is at these points where the assemblage seems most like a solid structure. Even in this nearly immobile space/time, territories or structures of normalized meaning can potentially act together to produce a possibility for discursive disruption. In chapter two, for instance, the clash between the highly classified and framed discursive and spatial constraints within the tattoo convention and competition produced a threshold space/time that made the rematerialization of the tattooed body possible. Specific, small, and possibly mundane movements of tattoo contests entrants (a small smile, a wiggle of a belly) were transformed in this clash of territories into moments of disruption and transformation.

In addition to occasions where glacial territories collide, space/time relations were also enacted within the processes and mechanisms through which discourse made and contained the meaning of the tattooed body. One of those mechanisms performed the spatializing process of physical and discursive fragmentation: taking the body apart and reassembling it in order to control its meaning. Despite the force of this maneuver, time, as expressed through movement of and within the fragmented parts, refused to be still. In chapter three, for example, representations in advertising contexts exposed the power of fragmentation that leaves the body open to rupture. This occurred in two related ways. In the first case, the analysis showed how the process of reassembling the body parts can expose the insufficiency of fragmentation to control the body's meaning. The seams and joins (scars, if you will) created in the wake of the discursive momentum to create a synthesized unity of these fragmented parts exposed the impossibility of that integration. The figures for the farce of cohesion were classic visualizations of the Frankenstein monster or the cracked shell of the fallen Humpty Dumpty. Secondly, the analysis suggested that in the process of taking the body apart, gendered meaning becomes attached to individual fragments (a hand or an arm, for instance). Upon reassembly, the parts stubbornly refuse to give up their individuality and instead try to inject their meaning into the new context (masculine arm placed upon female torso). Like the dye injected into the skin during tattooing, the meanings remain visible and traceable rather than absorbed and assimilated.

Folding emerged as the third space/time relation to promote image agency. "Fold" is, of course, both noun and verb. As an action, the process of folding requires the manipulation of already stratified structures. In straightforward terms, one must have something, some pliable object, to fold. In chapter four, the discourse of modern primitivism was folded or wrapped around the images and lives of Tattoo Mike and Captain Don in order to hide their distinctiveness. The activity of folding attempted to obscure difference but, as the analysis illustrated, actually generated the capacity of the hybrid. Once ensconced within the folded discourse of modern primitivism, Mike and Don materialized as representatives of the insufficiency of the discourse to contain their now both working-class and modern primitive hybridity. Their presence irritates modern primitive discourse. It compels us to tug at its corners or scratch at its surface. The hybrid, therefore, disrupts the force of the object of the fold by compelling or inciting unfolding or unraveling. Time (movement) again disrupts space (territory).

Despite the important and sometimes subtle differences in the agencies that emerged through each of the representational contexts considered in this analysis, another running theme (in addition to space/time relations) surfaced. That theme has to do with the power manifested in, and the alterations in meaning made possible by, the relations among, between, and through parts, fragments, and territories. In each of the cases considered, bodies of human beings, representations, and discourse were, in some fashion, dismembered. The resulting pieces and fragments (i.e., body parts, class identifications, tattoo images) collided against each other, invaded each other's space, contaminated each other, and were hidden in the crinkles and creases of discourse. In each instance the collisions, invasions, contaminations, and creases either disrupted the normalizing force of discourse or, at least, demonstrated the potential to do so. I characterize these relations among parts and fragments variously as thresholds, injections, infections, and foldings. What unites these different styles of engagement and suggests their importance to image agency is that they require some level of stratification or territorialization for their very existence. Their power to change and transform discourse and meaning results *from* forces of containment and repression.

For instance, in the case of the campaign for emergency contraception described in the conclusion to chapter three, the possibility for disrupting the gender constraints of the masculine arm tattoo of the pierced heart occurred because the pierced heart was injected with the symbol of emergency contraception. It is because emergency contraception, the arm tattoo, and the pierced heart are all so concretely gendered—so completely captured as traces—that the injection of one into the other created the disruptive possibility for the image to act out. At the tattoo convention, the forces of containment materialized in solid territorialized spatial and discursive boundaries. Agency in the form of the threshold emerged in the collision between discursive class and performance norms through the movement of contest entrants, and between spatial territories in the interstice linking the interior of the convention to the exterior hotel lobby. Again, it is because of the strength of the territorialized boundaries that the point at which they collide became energized with the transformative power of the threshold. Finally, modern primitive representations created the hybrid power of the disidentified working-class bodies of Mike and Don because their highly structured and territorialized class identities withstood the discursive forces that attempted to fold them into another discursive (modern primitive) territory.

In all of these contexts, the particular agencies were the result of containment and capture and not of infinite openness and unconstrained possibility. As I have been suggesting throughout this project, these examples show that agency can be productively tied to representational reproduction. *Agency is*, to reiterate from chapter one, *the enactment of the potential for disruption, destabilization, transformation, and change that is produced in and through reiterative, reproductive, and citational visual and discursive practices of meaning-making and meaning-attribution.* The value in recognizing this potential within extremely normalizing discourses is both in perceiving the agency of the image, as well as possibly perceiving agency more widely as made possible by and through representational constraints. This is not to say that the version of agency operating within the performance of performativity eclipses or is more important or valuable than forms of agency that depend on autonomous subjects acting in response or resistance to oppression. Nor is it to reject understandings of agency that rely on transgressive behaviors (both embodied and discursive). Rather, it is to say that the movements and disruptions occurring within the structures of discursive reproduction, within the interpretation of meaning, offer yet another opportunity for change and transformation. Within the contexts and images discussed in this project, those disruptions occurred through and in collisions of meaning, seams and sutures of fragments, infections, and folds. These are not the only ways through which the performance of performativity opens up into disruption and transformation, but they do provide some insight into how this process of agency works.

Critical Agency

Throughout the analysis of agency and the visual, I have become increasingly aware that in order to perceive these moments of agentic potential, maintaining the subjunctive posture of the "as if" is critical. In order to perceive the moments when visual images acted out of the repetitious and normative meaning-making processes, it has been important to follow the paths of Deleuze and Mitchell and approach the image "as if" it were alive. This approach was often a challenge to which this author did not always quite live up. Despite the analyses' important values of articulating and facilitating an understanding of the force of the visual within discursive structures, they also occasionally stop short of completely living up to the life of the image. As a result, readers may have been occasionally disappointed when the analysis of a particular image or context seemed to not quite fully evoke an image's potential. Other readers may also have perceived something else going on with the images that the analysis sidestepped, glossed over, simply missed, or blatantly ignored.

One partial explanation for this is that my primary purpose throughout this book has been to follow the images after the fact in order to perceive their already patterned movements in order to learn something about the possibilities for agency enacted within various discursive, visual, and material contexts. *Post hoc* analyses of this type are valuable because they identify and expose the processes and strategies through which agency emerges within the performance of performativity. They also provide, at least in this case, an articulation of specific figures (thresholds, folds, hybridity) that may be the impetus for future critical encounters that engage the image both after and *as* it moves. I imagine that another explanation for the analyses occasionally falling just short of realizing the complete potential of the image is due to the style of my critical engagement—that is, linear progression and argument. The mode of engagement that I find most comfortable and safe is one that relies on and reproduces the discipline of linearity. If, consequently, this pattern of critical relation fails to energize the specific character of the images' agency within any given discursive framework, then the criticism threatens to participate in reterritorializing the image. Instead of realizing the images' potential, the critical enterprise can end up reinscribing and/or prefiguring meaning.

Rather than taking these moments of near misses as critical mistakes, however, I want to use them as inspiration to speculate about the critic/audience/viewer's role in igniting the immanent potential of images to act in the process of criticism/audiencing/viewing. In my own incomplete and tentative way, I am heeding Rebecca Schneider's call to explore the productivity of my critical infelicity.[1] This speculation begins with a brief discussion of some relatively recent developments in the practices of performance, cultural, and literary criticism and concludes by bringing us back to the tattooed body for some additional critical provocations.

PERFORMATIVE WRITING/CREATIVE CRITICISM

The role of critical method in the humanities in general and in aesthetic culture (literary, visual) and performance in particular has recently been the focus of some important discussion.[2] In the Introduction to the edited collection *After Criticism*, for instance, Gavin Butt suggests that "criticism is in trouble."[3] As I suggest in chapter one, Hans Ulrich Gumbrecht has challenged the practice of meaning interpretation as "the exclusive core practice" of the humanities.[4] Moreover, it is the dedication to diagnosing meaning and the limitations in that maneuver that motivated Lawrence Grossberg to request criticism that goes elsewhere. In conceptualizing alternatives to the interpretation of meaning, the route has often been to aim for writing practices that evoke the experience and/or embodiment of spectatorship and reading. The solution to the problem of meaning has been recourse to the epistemology of the body. For instance, in his discussion of a "poetics of liveness," Matthew Reason argues for modes of writing and documenting live performance that communicate the "experience of performance."[5] Drawing from Della Pollock's influential description of performing writing, Reason concludes that the process of documentation should evoke both the aesthetics of the performance event as well as the embodied experience of the spectator.[6] Similarly, in the realm of literary criticism, Gerry Brenner suggests that the value of "performative criticism" is to participate in the "experience of the text critics are writing about" by using style in order to encourage the audience of criticism to become "as emotionally engaged as when it interacts with a good literary work."[7] In each of these cases, meaning is shoved aside in favor of an affective or physical response.

Concurrent to the discussion of the method in the literary and art enterprises, there has been a conversation in the subfield of performance within the discipline of Communication Studies surrounding notions of " 'personal narrative,' 'autobiographical performance,' 'mystory,' 'performance art,' 'poetic scholarship,' 'performative writing,' 'autoethnography,' and whatever we will have called it tomorrow, when that tomorrow becomes yesterday."[8] Performative writing, in the most general sense, represents a variety of writing forms and practices designed to expand the ways in which language performs in/as scholarship. The importance of embodied experience, as well as aesthetic style, also runs throughout this discussion.[9] For instance, descriptions of performative writing note that its practice and form integrates individual, personal, subjective, and embodied experience into academic prose.[10] Other accounts emphasize performative writing's heightened attentiveness to the aesthetic and creative capacities of language.[11] Still others suggest that performative writers experiment with a variety of styles and forms of representation.[12] Sometimes the writing appears autobiographical. Occasionally there is an overt manipulation of visual form ("fun with fonts," as a good friend of mine cheerfully opines). On other occasions, there is the practice of adopting literary genre (drama, poetry) as a style of presentation. The conceptual foundation for much of this work is located in the complementary critical desires to resist oppression of the (cultural, embodied, textual) other through colonizing representational practices and to provide a voice for subjugated knowledge, experience, and people. It evokes, in Ronald Pelias's terms, the often (in scholarly writing) elided and hidden openness of speaking and writing from the heart.[13]

I completely agree with and support the need and value of highlighting and integrating embodied experience through aesthetic style in the documentation and analysis of visual culture. I am further inclined to pursue yet another (perhaps perpendicular, or maybe parallel) route. This inclination is borne out of a desire to disrupt dialectical structures that draw lines between experience and meaning, embodiment and interpretation.[14] I certainly appreciate the importance of evoking the aesthetic and experience of the performance, literature, and /or art object. I think, however, that there are also alternative routes to critical agency that do not necessarily rely upon using experience as the antidote to objectivity or embodiment as the other to interpretation. In the same way that I am suspicious of the tendency in tattoo scholarship to flatten out the complexity of the process, practice, and representation of tattooing through the limiting categories of the inscribed or resisting subject, I am also hesitant to rely on embodiment or experience as the primary impetus behind performative writing/criticism. The route that I am interested in, as I suggest throughout the book, delves into the space and time of meaning's performative (re)production. The contribution of the analyses in this volume have been consequently to push critical attention through meaning interpretation into the performance of meaning's reproduction. The following foray into describing some possible performative critical encounters based on the performance of performativity and the practice of tattooing is inspired by Gingrich-Philbrook's implicit call for more metamethodological discourse about performative writing practices.[15] I am, in short, about to proceed to do much more abstract telling and describing than concrete showing or doing. Before considering some possibilities for how tattooing and the figures for agency that emerged throughout the analyses in this book may offer provocations for performative criticism/writing, let me offer a few more comments about what I perceive are some of the values and pitfalls of performative criticism to a project invested in the agency of the visual.

As for the opaqueness of some performative writing, the look on some of my student's faces is enough to cause some concern. I have on more than one occasion offered samples of performative writing or criticism to students and colleagues as examples of innovative work of and from performance studies. These samples vary in their topics and disciplinary sources but tend to be united in their challenge to the veneer of objectivity, linear argument, and classic essay organization. They actively resist allowing the reader to fall into an easily scripted relation to the critical text (looking for the thesis, perceiving paragraph structure) because the writing does not tell them what it means—it attempts to show them. Sometimes folks "get" it. Often, they do not. When they don't, the responses range from outright confusion to immediate dismissal. In some cases, the dismissal is a response to the confusion. The reader rejects the writing for its difficulty or lack of clarity.[16] In other cases, the rejection is conditioned by evaluating the writing through an already-held script for a genre of writing other than the critical essay—judging the performative essay, for example, that adopts the stance of personal narrative prose as being merely therapeutic expression or an uncritical autobiography. Gingrich-Philbrook articulates this tension as being the double bind of the epistemic/aesthetic—that is, the dual function of, in this context, criticism to both serve and create knowledge and to do so artfully without reifying ideological structures that privilege certain forms of knowledge. It is this struggle that spurs both the fanatical support of performative practices as well as its most vehement critics.

When it is at its best, performative writing navigates the problem of opaqueness and the epistemic/aesthetic double-bind by encouraging critics to attend to scholarship as an activity of aesthetic engagement that is highly contextual, evocative, and nervous.[17] It is contextual in its consideration and commitment to the genealogy of the image/performance/text, to traditions of critical knowledge, and to both the aesthetics of writing and of the text (or performance or art) critically engaged. It evokes the possibilities and potential of the subjunctive critical enterprise; the "as if" of the image. It is nervous about reifying "experience" and the uncritical "I" as ideological systems by keeping them present but, in Mary Poovey's terms, on the edge in "peripheral vision."[18] It is also nervous about its ethics. It remains answerable to its audience and object, context and history, concept and text, self and other. Finally, and particularly relevant to the agencies described herein, the mode of engagement is one of entering into the practices of critical encounter with an attention to the relations of time and space both in response to and within interpretations and structures of meaning. Sometimes, as in the case of most of the analysis encompassed in this text, this writing will perform traditionally. In other cases, the engagements with the text will spark other movements of inventive practice. The impetus behind the essays collected in *After Criticism* is, for example, to explore the critical performative possibilities of troubling conventional narratives of history, of engaging in modes of interaction with aesthetic objects that are distracted or disinterested instead of evaluative and judgmental, and of emphasizing critical writing as being both open to experimental form as well as a being a creative process of critical invention where ideas, concepts, and innovations in understanding are made to happen.

But What about Tattoos?

In the Introduction I suggested that the tattooed body was an important sight/site for the analysis of agency and visual images both because of its metaphoric and literal weight. Throughout the pages of this volume I have tried to be true to the genealogy of the performance and public display of the tattooed body and the ways in which that history has been represented and reinvented both in verbal and visual discourse. From the stages of P.T. Barnum and the disrobing of Captain Costentenus and Irene Woodward, to the midways of world's fairs and circuses, to the streets of New York and Chicago in the mid-twentieth century, to the pages of illustrated books and popular women's magazines, to contemporary advertising, I have traced the multiple surrogations of tattooed bodies and spaces. In the process, I have come to recognize the pivotal importance of critical orientation to the practice and ethics of criticism. This is partly due to my ongoing investment throughout the process in the performances of tattooed bodies and in the performativity of the performances of images and meanings. It is also partly due to the sheer force of the tattooed body itself. It is at this point, then, that I want to return to tattoos and the tattooed body as a source of inspiration for future, perhaps more creatively and aesthetically engaged, critical enterprises as well as to ask (but most likely not answer) again what makes the tattooed body such a lively and seemingly constant source of cultural fascination in the United States despite nearly 150 years of its history of public display.

Despite the narrative of deviance and resistance versus recuperation that, as I demonstrate in chapter one, drives most of the current and past evaluations of tattoos and the tattooed body, there are some alternative discourses within the tattoo community that can inspire critical engagements that attend to the performance of meaning's performativity (and to the variety of injections, infections, and foldings) by accomplishing more than diagnosis and description. The first has to do with the concept of "reworking" a tattoo. Many of the photo albums that are scattered on display booth tables at tattoo conventions are filled with "before" and "after" pictures of tattoos that have been altered and improved by the tattoo artist. Tattoos artists are often asked to rework or alter a tattoo produced in either the recent or distant past. In these cases, the tattoo artist performing the image alteration creates a new image in collaboration with the existing tattoo and, by extension, the tattooed individual and the (possibly unknown) original artist. If the "original" image was based on tattoo flash, the community of artists responsible for the tattoo expands even more because flash art is produced by and sold to tattoo artists around the world. The tattoo is, in the case of reworking, inherently an unfinished text produced collaboratively over and through time. In fact, the practices of getting tattooed by multiple artists, of artists reworking old tattoos, and of replicating and mass distributing flash art are so commonplace that the notion of a purely "original" and static tattoo text or body is rare, indeed.

What is more, there are many motivations for getting a tattoo redone. Perhaps the ink has faded over time. Perhaps the skin was injured at the site of the tattoo and the image marked or marred in some way. Perhaps the individual wants to expand the tattoo and thus extend the canvas of whatever body part is in question. Perhaps a new color or more detail is desired. Perhaps the event that the tattoo symbolizes has changed and so the tattoo needs to also change. Perhaps the alteration does not touch the original image but adds another image near it, thus remaking the context of the display. The skin changes, the tattoo fades, and another artist is commissioned to work on the text to create something new. The artist refreshes faded ink so that the lines of the image are clear again, highlights or repairs damage, adds new elements that radically alter the image while retaining vestiges (sometimes more than others) of the original text. Sometimes the effect is subtle, sometimes overstated, sometimes ironic, occasionally parodic (as in when Johnny Depp had his "winona" tattoo truncated to read "wino").

Each of these processes of alteration offers clues to sophisticated artist/critic/text relations. In fact, the ethic of reworking a tattoo is very much like treating images as if they were alive—perhaps even more so because, in a very material way, tattoos are living images because their canvas is the skin. Performative criticism modeled from tattoo reworking would attend to the aesthetics and content of the performance, visual, verbal, and/or literary text while at the same time taking what has been perceived and learned and then further reworking it. The practices available for the critic to refresh, highlight, repair, add, and alter images (as suggested by the lessons learned from images of the tattooed body) include processes of infection, reassembly, and injection.[19] If, for instance, I were to take my own advice and rework Rosie to be faithful to the life of her image, I would adapt what I have learned about the performance of her performativity in order to inject new images with her fragmented body parts infected with gendered meanings. Furthermore, tattoo artists and, by metaphoric extension, performative critics (the good ones) are respectful of the bodies and images that they are reworking while at the same time exploring ways in which the image can be made anew. Creative criticism in this mode would be respectful of the traditions and histories encountered in the process of critique and invention.

Occasionally, tattoo artists are asked to completely cover up a tattoo. The aim is to hide or eliminate an image by overinscribing it with another. Despite the fact that the artists' photo albums show some amazingly successful cover-ups in "before" and "after" shots, most will readily admit that the covering tattoo nearly never fully hides the original image—at least not to the critically trained eye. The unicorn head that I had inscribed on my shoulder in the late spring of 1991 was covered up at Tattoo Voodoo in 2000. As I consulted with the tattoo artist, he indicated that the image I had chosen would work to cover the unicorn, with a few alterations. The center of the image needed to begin with a layer of dark (purple or black) ink so that it would disguise most of the small unicorn (also in black ink). The size, shape, and color of the unicorn consequently partially dictated those of the replacement tattoo. Tattoos cannot be completely erased by other tattoos (or even by laser removal—there is always a scar, a remainder), they can merely be partially hidden under another layer of ink. Furthermore, the memory of the first tattoo remains in the form of the subsequent tattoo—both shaping the newer image as well as lurking in its shadows.

As a result, artists performing cover ups are almost as invested in the specifics of the first tattoo (albeit differently) as are those performing rework. Even in the case of cover ups, therefore, the act is collaborative. For instance, my unicorn was a complete mess. It was a disaster because the tattoo itself was not very good to begin with, because I mistreated the tattoo with excessive sun exposure, and because my skin aged. Despite the fact that the unicorn was a wreck—a flash-inspired, poorly rendered, ugly, faded, and misrecognized image—the artist charged with concealing it had to pay attention to it. He could not simply pretend it did not exist. My little unicorn just would not be ignored. The unicorn stubbornly influenced his (perhaps) more sophisticated aesthetic vision. This process of concealment and collaboration over time describes one version of surrogation and the transmission of cultural memory. It also suggests that critical impulses can be directed toward the unsophisticated and ugly and need not always strive for simple clarity and transparency. I am not endorsing writing practices that are completely inaccessible to the audience—that is, they are too successful in obscuring the stakes of their construction. Rather, I am suggesting that it might be occasionally productive to participate in some crafty covering up—as long as the occluding image retains the echo of the original. The performative critic would then ask what new meaning or aesthetic text can be made out of some ugly (appropriative or oppressive) cultural artifact or text. The caveat again, however, is that the new performance/image/text made should attend to the history and aesthetic of the performance/image/text remade or covered up.

It seems clear to me that the genealogy of the tattooed body's performance, the practice of tattooing, the actions of tattoos, and the *performativity* of meaning offer a much more sophisticated and complex version of criticism, tattooing, and visual agency than do basic attributions of deviant, gender, class, or exotic meaning. The tattoo is always in motion, is rarely a uniquely individual expression, is usually waiting for change and alteration, and is not nearly as "permanent" as we would have it. It is, perhaps, through these capacities of the tattoo that we can explain the continuing fascination with tattooing. Not only because tattoos mark the body in possibly unnatural ways that may violate normative standards of beauty and not only because tattoos have a conflicted history of marking deviance, but also because tattoos, despite appearances to the contrary, never quite stand still. They remind us of the always immanent potential for change lying even within the most permanent-seeming territories.

Notes

Introduction: Picturing Tattoos

1. Gilles Deleuze and Felix Guattari, *A Thousand Plateaus* (Minneapolis, MN: University of Minnesota Press, 1987), 86.
2. In this way, the "trouble" that my image causes is related to, albeit slightly different than, the way in which Judith Butler mobilizes the term trouble. Judith Butler, *Gender Trouble* (New York: Routledge, 1990).
3. Joseph Roach, "Culture and Performance in the Circum-Atlantic World," in *Performativity and Performance*, ed. Andrew Parker and Eve Kosofsky Sedgwick (New York: Routledge, 1995), 48.
4. Joseph Roach, *Cities of the Dead: Circum-Atlantic Performance* (New York: Columbia University Press, 1996), 11.
5. Ibid., 12.
6. Ibid., 2.
7. It is important to note here that I am not necessarily referring to linear notion of history. "Prior" is a construction or fiction that is enabled through the process of surrogation and the actual citational links between and among cultural events occur along multiple pathways and in multiple directions. A rhizomatic distribution, if you will.
8. Roach, *Cities of the Dead*, 3.
9. Chapter One provides an extended discussion of the association between tattoos and deviant behavior.
10. Roach, *Cities of the Dead*, 27.
11. Ibid.
12. Ibid.
13. Ibid.
14. Roach, *Cities of the Dead*, 27–28.
15. W.J.T. Mitchell, *What do Pictures Want* (Chicago, IL: University of Chicago Press, 2005), xiii.
16. See, for example, Paul DuGay, Stuart Hall, Linda Janes, Hugh Mackay, and Keith Negus, *Doing Cultural Studies* (London: Sage, 1997).

Chapter One Vital Images: The Performance of Performativity

1. Stuart Hall, "Cultural Studies and its Theoretical Legacies," in *The Cultural Studies Reader*, ed. Simon During (New York: Routledge, 1999), 107.
2. Dick Hebdige, *Subculture: The Meaning of Style* (London: Methuen, 1979), 139.
3. Peggy Phelan, *Unmarked: The Politics of Performance* (New York: Routledge, 1993), 2.
4. William DeMichele, *The Illustrated Woman* (New York: Proteus Press, 1992), 73.
5. Judith Butler, *Bodies that Matter* (New York: Routledge, 1993), 10.
6. Citationality refers to the ways in which representations and/or performances explicitly or implicitly refer to or depend upon prior and concurrent representations for their meaning. Jennifer's picture, for example, could be argued to cite the conventions of wanted posters with the close-up shot of the face centered in the frame. Alternately, the image also arguably refers to other popular cultural images of women with shaved heads. For instance, the photo was taken and the book published in the early 1990s, when Sinéad O'Connor was at the height of her popularity after the release of *I Do Not Want What I Haven't Got.* O'Connor's famously shaved head and her subsequent political protest—such as tearing apart the image of the Pope John Paul II on *Saturday Night Live* (October 3, 1992) in protest of Catholic priest sexual misconduct—are consequently part of the citational story of Jennifer's image.
7. Ott has recently suggested, for example, that the search for representational resistance and normative reproduction has become predictable and boring. Brian Ott, "(Re)Locating Pleasure in Media Studies: Toward an Erotics of Reading," *Communication and Critical/Cultural Studies*, 1 (2004): 194–212. Wiley also indicates that despite Lawrence Grossberg's call for an alternative approach to issues of politics and agency in cultural studies, that there is a continued modernist assumption behind cultural critiques of representation. Stephen Crofts Wiley, "Spatial Materialism: Grossberg's Deleuzian Cultural Studies," *Cultural Studies*, 19 (January 2005): 63–99.
8. Wiley, 65. I do not mean to say, however, that there are not critiques of this type of representational economy. Clearly, the reliance on the work of Deleuze and Guattari in this book indicates that the epistemology of representation has been long under fire.
9. Hans Ulrich Gumbrecht, *Production of Presence: What Meaning Cannot Convey* (Stanford, CA: Stanford University Press, 2004), 1.
10. Lawrence Grossberg, "The Victory of Culture: Part One," *Angelaki: Journal of Theoretical Humanities*, 3.3 (1998): 12, 16.
11. Wiley describes this as the logical structure of identity politics: "which conflates identity, subjectivity, and agency and assumes that all three coincide in individual or collective actors defined by logics of difference" (65).
12. In a recent essay, Kosut suggests that there will not soon be an end to the association between tattooing and deviance and that this has relatively permanently

marked tattooing as a fringe art practice associated with, in Kosut's argument, asylum art. Mary Kosut, "Mad Artists and Tattooed Perverts: Deviant Discourse and the Social Construction of Cultural Categories," *Deviant Behavior*, 27 (2006): 73–95.

13. Mark Gustafson, "Curiously Marked: Tattooing and Gender Difference in Eighteenth-Century British Perceptions of the South Pacific," in *Written on the Body*, ed. Jane Caplan (Princeton, NJ: Princeton University Press, 2000), 24.
14. Ibid., 29.
15. Charles W. MacQuarrie, "Insular Celtic Tattooing: History, Myth and Metaphor," in *Written on the Body*, ed. Jane Caplan (Princeton, NJ: Princeton University Press, 2000), 44.
16. Ibid., 33.
17. Hamish Maxwell-Stuart and Ian Duffield, "Skin Deep Devotions: Religious Tattoos and Convict Transportation in Australia," in *Written on the Body*, ed. Jane Caplan (Princeton, NJ: Princeton University Press, 2000), 118–135.
18. Abby M. Schrader, "Branding the Other/Tattooing the Self: Bodily Inscription among Convicts in Russia and the Soviet Union," in *Written on the Body*, ed. Jane Caplan (Princeton, NJ: Princeton University Press, 2000), 175. Anderson's discussion of the use of tattoos as punishment in Indian penal colonies indicates, however, that this practice was not always appropriated as a means of resistance as Indian colonist sought to cover or erase the tattoos. Clare Anderson, "*Godna*: Inscribing Indian Convicts in the Nineteenth Century," in *Written on the Body*, ed. Jane Caplan (Princeton, NJ: Princeton University Press, 2000), 102–117.
19. Jane Caplan, "'National Tattooing': Traditions of Tattooing in Nineteenth-Century Europe," in *Written on the Body*, ed. Jane Caplan (Princeton, NJ: Princeton University Press, 2000), 156–173.
20. James Bradley, "Body Commodification? Class and Tattoos in Victorian Britain," in *Written on the Body*, ed. Jane Caplan (Princeton, NJ: Princeton University Press, 2000), 139.
21. Margo DeMello, *Bodies of Inscription: A Cultural history of the Modern Tattoo Community* (Durham, NC: Duke University Press, 2000); Samuel Steward, *Bad Boys and Tough Tattoos* (New York: Haworth Press, 1990); Clinton Sanders, *Customizing the Body* (Philadelphia, PA: Temple University Press, 1989); Elizabeth Seaton, "Profaned Bodies and Purloined Looks: The Prisoner's Tattoo and the Researcher's Gaze," *Journal of Communication Inquiry*, 11.2 (1987): 17–25.
22. Alan B. Govenar, "Introduction," in *Stoney Knows How: Life as a Tattoo Artist*, ed. Leonard St. Clair and Alan B. Govenar (Lexington, KY: University of Kentucky Press, 1981), xx.
23. Govenar, "Introduction," xx, xxi.
24. Ibid., xxi.
25. Although not always directly mentioned, it should come as no surprise that much of the theorization of the social inscription and resistance dynamic owes a debt to Foucault's famous work in *Discipline and Punish*. The

language of disciplinary power circulates though many of the arguments regarding the act of tattooing slaves and conflicts and their acts of resisting that inscriptive imposition. See Gustafson and Maxwell-Stewart and Duffield particularly. Also, Michel Foucault, *Discipline and Punish* (New York: Vintage, 1979).

26. See Gustav Newman, "The Implications of Tattooing in Prisoners," *Journal of Clinical Psychology*, 43 (1982): 231–234; Norman Goldstein, "Psychological Implications of Tattoos," *Journal of Dermatologic Surgery and Oncology*, 5 (1979): 883–888; A.J.W. Taylor, "Tattooing among Male and Female Offenders at Different Ages in Different Types of Institutions," *Genetic Psychology Monographs* 81 (1970): 81–119; J. Briggs, "Tattooing," *Medical Times* 87 (1958): 1030–1039.
27. Nikki Sullivan, *Tattooed Bodies: Subjectivity, Textuality, Ethics, and Pleasure* (Westport, CT: Praeger, 2001), 21.
28. Arnold Rubin, "The Tattoo Renaissance," in *Marks of Civilization: Artistic Transformations of the Human Body*, ed. Arnold Rubin (Los Angeles, CA: Museum of Cultural History, University of California, 1988): 233–264; Jane Caplan, "Introduction," in *Written on the Body*, ed. Jane Caplan (Princeton, NJ: Princeton University Press, 2000), xi; DeMello, *Bodies of Inscription*; Sanders, *Customizing the Body*, 18–20.
29. DeMello, *Bodies of Inscription*, 143–151.
30. Victoria Pitts, *In the Flesh: The Cultural Politics of Body Modification* (New York: Palgrave, 2003), 34.
31. Ibid., 35.
32. Ibid., 40.
33. TV.com, "Meet Chris Nuñez," *Miami Ink Video Gallery* (accessed July 28, 2005). http://www.tv.com/tracking/viewer.html?tid=13497&ref_id=37627&ref_type=101&om_act=convert&om_clk=headlinessh&tag=headlines;title;3.
34. Katherine Irwin, "Saints and Sinners: Elite Tattoo Collectors and Tattooists as Positive and Negative Deviants," *Sociological Spectrum*, 23 (2003): 27–57.
35. Irwin, 38–41.
36. For the idea about the fertility of reproduction, I am indebted to W.J.T. Mitchell.
37. Hariman and Lucaites use this sense of the term in their analyses of iconic photographic images, for example. See Robert Hariman and John Lucaites, "Performing Civic Identity: The Iconic Photograph of the Flag Raising on Iwo Jima," *Quarterly Journal of Speech* 88.4 (2002): 363–393.
38. For example, Amelia Jones's work on body art tends towards this deployment of the term. Jane Blocker's revisitation of Jones' work does so as well. Blocker refers, for example, to "performative artistic practice" (9). Amelia Jones, *Body Art: Performing the Subject* (Minneapolis, MN: University of Minnesota Press, 1998); Jane Blocker, *What the Body Cost: Desire, History, and Performance* (Minneapolis, MN: University of Minnesota Press, 2004).
39. Richard Bauman, for instance, defines performance as heightened states of communication or activity. Richard Bauman, "Performance," in *Folklore,*

Cultural Performances, and Popular Entertainments, ed. Richard Bauman (New York: Oxford University Press, 1992): 41–49.

40. J.L. Austin, "Lecture I," in *How to Do Things With Words*, ed. J.O Urmonson and Marina Sbisa (Cambridge, MA: Harvard University Press, 1975), 5.
41. Jacques Derrida, "Signature, Event, Context," in *Limited Inc.*, ed. Gerald Graff (Evanston, IL: Northwestern University Press, 1988), 17.
42. Butler, *Bodies that Matter*, 2.
43. Ibid., 12–13.
44. Ibid., 12.
45. Erving Goffman, *Presentation of Self in Everyday Life* (Garden City, NY: Doubleday, 1959).
46. Marvin Carlson, *Performance: A Critical Introduction* (New York: Routledge, 1996), 3.
47. See, for instance, the ethnographic work of Dwight Conquergood, Joni Jones, Soyini Madison and others.
48. In using this particular phrase, I am alluding to *By Means of Performance*, ed. Richard Schechner and Willa Appel (Cambridge, UK: Cambridge University Press, 1990).
49. Della Pollock, "Introduction," in *Exceptional Spaces: Essays in Performance and History*, ed. Della Pollock (Chapel Hill, NC: University of North Carolina Press, 1998), 22.
50. Richard Schechner famously defines performance as restored behavior. Performance is always a restoration of a (part of) a past behavior or series of behaviors. See Richard Schechner, *Between Theater and Anthropology* (Philadelphia, PA: University of Pennsylvania Press, 1985).
51. Within academic discussions of agency and body representation, for example, it has been the extraordinary and/or excessive practices that have garnered the most attention. Performance artists and posthuman cyborgs that hyperbolize confrontations with normative discourses tend to be thoroughly discussed whereas the more everyday and mundane acts are overlooked. Exceptions include Janice Radway, *Reading the Romance* (Chapel Hill, NC: University of North Carolina Press, 1984).
52. Mitchell, *What Do Pictures Want*, 90.
53. Mitchell argues throughout *What Do Pictures Want* that the second commandment's prohibition of creating images in the likeness of God provides the cultural backing for the vital power of images (133–134).
54. Mitchell, *What Do Pictures Want*, 31. My interpretation here differs somewhat from Mitchell's conclusion. Despite the comparison that Mitchell makes between images and viruses (89), Mitchell also suggests that the legs of the image imply an "intelligence and purposiveness of their own" (31). In short, he concludes that, at least metaphorically, the image is an autonomous actor. As an alternative, I am offering the notion of viral reproducibility as the form of life of the image.
55. Mitchell, *What Do Pictures Want*, 46.
56. Ibid., 351. Another notable example is in Mitchell's discussion of the ventriloquist-like attribution of speech acts to images (140).

57. Mitchell, *What Do Pictures Want*, 351.
58. Henry Sussman, "Deterritorializing the Text: Flow-Theory and Deconstruction," *MLN*, 115 (2000): 974–966. Sussman suggests, "Deleuze and Guatarri insist that we take their projections of a highly privatized and singular conceptual system, which is always a potential one, a theory of the as-if, with the utter seriousness with which they do" (977).
59. I am referring here specifically to Baudrillard's formulation of the simulacrum, but one could easily also call upon others. Jean Buadrillard, "Simulacra and Simulations," in *Selected Writings*, ed. Mark Poster (Stanford, CA: Stanford University Press, 1988): 166–184.
60. Kevin Deluca, *Image Politics* (New York: The Guilford Press, 1999). Deluca uses the phrase "image event" to refer to the media tactics of environmental and other social protest movements. Deluca's analysis suggests that these events are examples of rhetorical expression that combines the discursive with the nondiscursive and the emotional with the rational in order to effect "critique through spectacle" (22). The cartoons, from the perspective of a group manipulating or using images to perform social critique, become an image event due partly to the Muslim cleric's act of delivering the images from Denmark to various church leaders in the Middle East. What distinguishes my approach to agency from Deluca's is the question of use or intent. Agency, in the terms of this project, is not limited to an individual or group using an image or images for any specific purpose.
61. Tom Conley, "From Multiplicities to Folds: On Style and Form in Deleuze." *The South Atlantic Quarterly*, 96.3 (Summer 1997): 629–646, 630.
62. Deleuze and Guattari, *A Thousand Plateaus*, 474–500.
63. Ibid.,504–505.
64. Ibid., 474.
65. This is, of course, Deleuze and Guattari's critique of a version of psychoanalysis or linguistics that places a structure upon behavior in such a way that (1) does not have to attend to the specificities of the action (2) by necessity creates moments of excess or exclusion which are either brushed off as anomalous or integrated Borg-like into the collective, and (3) guarantees its own status by always finding that which it is looking for.
66. One such mechanism is the process of signification or the "semiotic machine" (*A Thousand Plateaus*, 83).
67. See Michel Foucault, *Discipline and Punish*.
68. For a thorough discussion of the structure and effects of this tattoo discourse see Sullivan, 13–46.
69. I am borrowing the phrase "stop-operation" from Brian Massumi, *Parables for the Virtual* (Durham, NC: Duke University Press, 2002), 7.
70. Jonathan Crary, *Techniques of the Observer: On Vision and Modernity in the 19th Century* (Boston, MA: MIT Press, 1992), 31.
71. J. Macgregor Wise, "Assemblage," in *Gilles Deleuze: Key Concepts*, ed. Charles J. Stivale (Ithaca, NY: McGill-Queen's University Press, 2005), 77.
72. Deleuze and Guattari, *A Thousand Plateaus*, 4.
73. Wise, "Assemblage," 86.

74. Wise refers to the flow of agency in his discussion of assemblage (84). As I indicate earlier, however, these flows tend to be of deterritorializing lines of flight.
75. Wise, "Assemblage," 77.
76. Deleuze and Guattari, *A Thousand Plateaus*, 83, 87.
77. Ibid., 23.
78. Gilles Deleuze, *The Fold: Leibniz and the Baroque*, trans. Tom Conley (Minneapolis, MN: University of Minnesota Press, 1993), 20.
79. Deleuze and Guattari, *A Thousand Plateaus*, 506.
80. I think that move is warranted because Deleuze and Guattari insist that territorialization and deterritorialization not be reduced to a bad/good dualism. One interpretation of Deleuze and Guattari might suggest, for example, that agency (political resistance to power structures) is located in the activity of deterritorialization. The critical (read political) enterprise would emphasize the heightened movement of structures breaking down. This version of agency might privilege cultural, social, and artistic activities' capacities to question and destroy structure. This version of agency merely echoes the types of analysis of tattooing that have ended up stuck in the both/and economy of representation, however. As an alternative, therefore, I want to use this book to explore the possibilities for agency located in territorialization.
81. Deleuze and Guattari do, of course, discuss performativity thoroughly in the fourth plateau: "November 20, 1923: Postulates of Linguistics." Their discussion suggests that the illocutionary force of the performative utterance does not come from the subject or from intention because the act of the performative is to in fact produce the field of subjectivity. Their primary target is the relation between the statement and the act, arguing that this is an internal relation of redundancy. My concern is less with the relation between statement and act than with the activity of meaning itself—as I have stated before, with the performance of performativity.
82. Deleuze and Guattari, *A Thousand Plateaus*, 21.
83. Deleuze, *The Fold*.
84. Conley, "From Multiplicities to Folds," 639.
85. Deleuze and Guattari, *A Thousand Plateaus*, 13.
86. Ibid., 12.
87. Richard Schechner coined the phrase "restored behavior" as a definition of performance. This definition assumes that both aesthetic (theater, dance, opera, performance art) and everyday (social and cultural ritual) performances are actions, or scripts, that can be stored, transmitted, manipulated, and transformed. See Richard Schechner, *Between Theatre and Anthropology*.
88. Ott, "(Re)Locating Pleasure in Media Studies," 194–212.
89. Ibid., 207.
90. Deleuze and Guattari, *A Thousand Plateaus*, 13, 21.
91. Ibid., 493.
92. Ibid., 14.

Chapter Two Modern Day Sideshows: Class, Classification, and Tattoo Conventions

1. Portions of this chapter appeared in print previously as: "Movement and Resistance: (Tattooed) Bodies and Performance," *Communication and Critical/Cultural Studies*, 4.1 (March 2007): 51–73.
2. Pierre Bourdieu, "Distinction: A Social Critique of the Judgment of Taste," in *Social Stratification: Class, Race, and Gender in Sociological Perspective*, ed. David B. Grusky (Boulder, CO: Westview Press, 1994), 410.
3. John Fiske, *Reading the Popular* (Boston, MA: Unwin Hyman, 1989), 29.
4. Elizabeth Grosz, *Volatile Bodies* (Bloomington, IN: Indiana University Press, 1994), 139.
5. Barbara Ehrenreich, *Fear of Falling: The Inner Life of the Middle Class* (New York: Harper Perennial, 1989), 7.
6. Annalee Newitz and Mathew Wray, "What is 'White Trash'? Stereotypes and Economic Conditions of Poor Whites in the United States," in *Whiteness: A Critical Reader*, ed. Mike Hill (New York: New York University Press, 1997), 172.
7. Kaja Silverman, "Fragments of a Fashionable Discourse," in *On Fashion*, ed. Shari Benstock and Suzanne Ferriss (New Brunswick, NJ: Rutgers University Press, 1994), 189.
8. Grosz, *Volatile Bodies*, 138.
9. Tattoos can, of course, be removed or be hidden from view. The distinction, however, is that the tattoo on display is a sign signifying something more socially and physically implicating than clothing or hair styles. It is a sign that represents a choice to become permanently associated with a particular identity-an identity that during most of the twentieth century in the U.S. has been labeled criminal and deviant.
10. David Sibley, *Geographies of Exclusion* (New York: Routledge, 1995), 80.
11. Ibid., 81.
12. DeMello, *Bodies of Inscription*, 1, 26–32, 129–131.
13. Ibid., 21.
14. Madame Chinchilla, *Stewed, Screwed and Tattooed* (Fort Bragg, CA: Isadore Press, 1997), 62.
15. The term "tattoo collectors" refers to people who are not tattoo artists, have been tattooed more than once, probably intend to be tattooed again, are proud of their tattoos, and probably are not shy about showing their tattoos.
16. Chris Pfouts qtd in Amy Krakow, *The Total Tattoo Book* (New York: Warner Books Inc., 1994), 156.
17. DeMello, *Bodies of Inscription*, 28. See also Krakow, *Total Tattoo Book*, 154. The convention flyer advertised the tattoo competition as a "live" event. This did not, however, mean that tattooing would be practiced on stage. Rather, it meant that "live" bodies (as opposed to photographs) would be on display. The rhetoric of the "live" performance hinted clearly to the sideshow precursor. See

Carl Hammer and Gideon Basker, *Freak Show: Sideshow Banner Art* (San Francisco, CA: Chronicle Books, 1996).

18. DeMello, *Bodies of Inscription*, 29.
19. The distinction I am drawing here is between the competition as a specific event and the convention as a whole. People with tattoos primarily populate the convention and those tattoos are conspicuously displayed. While mingling on the convention floor, however, the attendees interact at multiple levels and the tattoo is not always the focal point. When on stage, however, the tattoos are the premier attraction.
20. The Expo returned to New Orleans in 2006 with a joint 2005/2006 event following the devastation of the city by Hurricane Katrina.
21. It is important to note that at this particular convention all of the bodies entered into competition were white bodies. Although I am unwilling to assert any definitive explanation for this, I would tentatively suggest that since tattooing as an artistic practice is dominated by white men and that tattoos on Hispanic, African American and Asian men in the United States are often associated with either celebrity status (athletes, musicians, actors, and models) or criminality (gang membership), that this setting had limited appeal. I would further suggest that the increased popularity of tattooing among the middle class is a largely white middle-class phenomenon.
22. Christine Braunberger, "Revolting Bodies: The Monster Beauty of Tattooed Women," *NWSA Journal*, 12.2 (2000): 1–23. Braunberger contests that in her experience at conventions, there was a distinction in the style of performance of the men and women on the tattoo competition stage. She indicates that the men seemed "grudging" to be on stage but that the women were not at all embarrassed to take the stage scantily clad in swimsuits (16). This was not the case at the Expo I attended. Nary a swimsuit did I see. Also, the relative level of grudgingness seemed more linked to personality and audience support than to gender. Men or women who were animated drew audience reaction and those who were not, did not. Perhaps the difference in my and Braunberger's experience is because the competitions Braunberger describes appear to award prizes primarily to the tattooed body whereas the Expo competition was clearly about the artist. She cites, for example, the comment of a "veteran tattoo collector" who felt she was not "feminine-looking enough" one year and so went on a diet and changed her costume and "won first place the following year" (17). I also think that the Southern Comfort Expo was not on as grand a scale as some of these other conventions and did not focus so exclusively on professional or artistic tattooing; it was a much more heterogeneous class mix. Finally, Southern Comfort is in the deep South and offers a very different cultural context than conventions held in California or New York.
23. Chinchilla, *Stewed, Screwed and Tattooed*, 62.
24. Krakow, *Total Tattoo Book*, 155.
25. Braunberger, "Revolting Bodies," 16.
26. Richard Bogdan, *Freak Show* (Chicago, IL: University of Chicago Press, 1988), 241–256.

27. The Human marvels traveled together from 1998–2005 when Enigma began performing solo. For more information about Enigma or Katzen, visit the Web site: http://www.humanmarvels.com/ (accessed February 1, 2007).
28. Hammer and Basker, *Sideshow Banner Art*, 36.
29. Bogdan, *Freak Show*, 242.
30. Ibid., 247.
31. Robert Sherwood, *Here We Are Again: Recollections of an Old Circus Clown* (Indianapolis, IN: The Bobbs-Merrill Company, 1926), 151.
32. Bogdan, *Freak Show*, 104–110.
33. Margot Mifflin, *Bodies of Subversion: A Secret History of Women and Tattoo* (New York: Juno Books, 1997), 12.
34. It was in the 1920s and 1930s when the tattooed performers lost their purely visual appeal because of a glut in the market. As a result, tattooed men and women began to perform acts like sword swallowing and becoming human "blockheads" in order to maintain their audience and, consequently, their livelihood.
35. Sherwood's (*Here We Are Again*, 149) account of Captain Costentenus' tattooing ordeal, for instance, indicates that it was punishment for his Christianity.
36. For this use of the term "dopplering," I am adapting from Brian Massumi's use of the term in discussing hypertext navigation/reading/use. Brian Massumi, *Parables for the Virtual: Movement, Affect, Sensation* (Durham, NC: Duke University Press, 2002), 138–140.
37. Butler, *Bodies that Matter*, 2.
38. It is also interesting to note that the contest assumes some distinction between male and female tattooed bodies. This distinction implies that tattoo design and style are gender coded.
39. Bogdan, *Freak Show*, 62–66.
40. Steve Gilbert, *Tattoo History: A Source Book* (New York: Juno Books, 2000), 141.
41. Ibid., 142.
42. Bogdan, *Freak Show*, 249.
43. Sibley, *Geographies of Exclusion*, 25.
44. Bogdan, *Freak Show*, 249.
45. Govenar, "Introduction," xviii.
46. For further reading about the power of the aesthetic gaze to finalize the form of the other see Mikhail Bakhtin, Author and Hero in Aesthetic Activity," *Art and Answerability: Early Philosophical Essays by M.M. Bakhtin*, trans. Vadim Liapunov and Kenneth Brostrom, ed. Michael Holquist and Vadim Liapunov (Austin, TX: University of Texas Press, 1990), 4–256.
47. Mary Douglas, *Purity and Danger* (Harmondsworth, UK: Penguin Books, 1970), 4.
48. Fiske argues, for example, that self-display is vulgar (29). See also Maureen Turim, "Seduction and Elegance: The New Woman of Fashion in Silent Cinema," in *On Fashion*, ed. Shari Benstock and Suzanne Ferriss (New Brunswick, NJ: Rutgers University Press, 1994), 140–158. Turim additionally

indicates that nudity in nickelodeon film was seen as a draw for low-class, but not middle-class, audiences (147).

49. Sanders, *Customizing the Body*, 18.
50. Ibid., 18.
51. Steward, *Bad Boys*, 102–127.
52. Ibid., 94–95.
53. Mifflin, *Bodies of Subversion*, 38; Krakow, *Total Tattoo Book*, 27.
54. Krakow, *Total Tattoo Book*, 28.
55. Tattooing was, for instance, just recently legalized in South Carolina (2004) and Oklahoma (2006).
56. Mifflin, *Bodies of Subversion*, 38.
57. Ibid., 148.
58. Timothy A. Roberts and Sheryl A. Ryan, "Tattooing and High-Risk Behavior in Adolescents," *Pediatrics*, 110.6 (December 2002): 1058.
59. Robert Raspa and John Cusack, "Psychiatric Implications of Tattoos," *American Family Physician* (May 1990): 1481.
60. Butler, *Bodies that Matter*, 10.
61. DeMello, *Bodies of Inscription*, 143–151.
62. Elmer Pace and Linda Pace v. City of Des Moines, Iowa and Brian Danner, United States Court of Appeals for the Eighth Circuit, U.S. App. LEXIS 3888, 2000.
63. United States of America v. Alfred Yazzie and Raymond Jones, United States Court of Appeals for the Tenth Circuit, U.S. App. LEXIS 18474, 1999.
64. DeMello, *Bodies of Inscription*, 5–6.
65. Braunberger, "Revolting Bodies," 16.
66. Sibley, *Geographies of Exclusion*, 80.
67. Bogdan, *Freak Show*, 46.
68. Hammer and Basker, *Sideshow Banner Art*, 14.
69. Bogdan, *Freak Show*, 100.
70. Leonard St. Clair and Alan Govenar, *Stoney Knows How: Life as a Tattoo Artist* (Lexington, KY: University of Kentucky Press, 1981), 32.
71. Bogdan, *Freak Show*, 101–102.
72. Hammer and Basker, *Sideshow Banner Art*, 19. The cardboard pirates also recollect the carnival banner.
73. Ibid., 10.
74. In its modern manifestation, however, authenticity is less the concern than artistic value, and the audience is a group of peers as opposed to outsiders.
75. Peter Jackson, *Maps of Meaning* (New York: Routledge, 1992), 91–96.
76. Neil Harris, *Cultural Excursions* (Chicago, IL: University of Chicago Press, 1990), 20.
77. Harris, 21–22.
78. Don Wilmeth, *Variety Entertainment and Other Amusements: A Reference Guide* (Westport, CT: Greenwood Press, 1982), 3.
79. Ibid., 95.
80. This is an interesting contrast to the mid-nineteenth century. Barnum's American Museum, for example, was located in the heart of New York at the

corner of Broadway and Ann. It was during the Jacksonian inspired 1840s that Barnum's museum flourished, however, when his audience was not as class segregated and the social attitude toward the display of human oddities was generally ambivalent.

81. Bogdan, *Freak Show*, 48.
82. Wilmeth, *Variety Entertainment*, 27.
83. Justus D. Doenecke, "Myths, Machines and Markets: The Columbian Exposition of 1893," *Journal of Popular Culture*, 7.3 (1973): 540.
84. Ibid., 542.
85. Wilmeth, *Variety Entertainment*, 22.
86. Bogdan, *Freak Show*, 58.
87. Although I think this is a weak explanation considering that tattoo conventions are an international affair. Tattoo artists from the United States travel from Tokyo to Amsterdam to attend conventions. At Tattoo Voodoo, for example, there were around fifty studios representing four countries including The Netherlands, Austria, Germany and the United States. American artists traveled from as far as New York, California, and Minnesota. Tattoo artists also charge a minimum of around fifty dollars for a tattoo that they can accomplish in forty-five minutes. In other words, tattoo artists are not necessarily people of small economic means.
88. The images from New Orleans during and after Hurricane Katrina both verify the ways in which the city was divided as well as suggest the disruptive power when those lines are crossed by nature and class. Seeing, for instance, the margins of society locked into the city center in the form of the black faces pleading for help at the Superdome and Convention Center demonstrated how the center can shift quickly to the margin when classed and raced bodies move from their prescribed places.
89. This area made international headlines in 2005 because it is also the site of the Astro Dome that served as the not-so-temporary housing for the thousands of evacuees form New Orleans during and after the devestation of Hurricane Katrina. The class implications of housing the influx of the low-income inner city population of New Orleans in the site of a popular amusement venue like a former sports stadium is not lost upon this author.
90. Kathy Peiss, *Cheap Amusements: Working Women and Leisure in Turn-of-the-Century New York* (Philadelphia, PA: Temple University Press, 1986), 95.
91. Christopher Gray, "Streetscapes/Roseland: An Old-Fashioned Dance to the Music of Time," *New York Times* (October 13, 1996). late edition, sec. 9: 5.
92. Peiss, *Cheap Amusements*, 94.
93. Elisabeth Perry, "Cleaning up the Dance Halls," *History Today* 39, 10 (October 1989): 22.
94. "Life Goes to a Party," *Life*, 2.5 (April12, 1937): 82–85.
95. Perry points out that even though the dime-a-dance or "taxi dance" women were not seen as prostitutes, they were associated with sexual disorder (25).
96. David Freeland, "Last Dance at the Orpheum," *New York Press*, 18.32 (2005), http://www.nypress.com/18/32/news&columns/DavidFreeland.cfm, par.9. (accessed January 25, 2007).

97. Bogdan, *Freak Show*, 30.
98. Patrick C. Easto and Marcello Truzzi, "The Carnival Social System," *Journal of Popular Culture*, 7.3 (1973): 555.
99. Ibid.
100. Ibid., 552; Bogdan, *Freak Show*, 10.
101. Easto and Truzzi, 559.
102. Mikhail Bakhtin, "Forms of Time and of the Chronotope in the Novel," *The Dialogic Imagination*, ed. Michael Holquist, trans. Caryl Emerson and Michael Holquist (Austin, TX: University of Texas Press, 1981), 84.
103. Ibid., 248.
104. Sibley, *Geographies of Exclusion*, 81.
105. Braunberger, "Revolting Bodies," 17.
106. Barbara Kirshenblatt-Gimblett, *Destination Culture: Tourism, Museums, and Heritage* (Berkley, CA: University of California Press, 1998), 55.
107. Jennifer DeVere Brody, "Black Cat Fever: Manifestations of Manet's *Olympia*," *Theatre Journal*, 53 (2001): 96.

Chapter Three Rosie Gets a Tattoo: Gender, Fragmentation, and Advertising

1. Considering the increasing popularity (or at least visibility) of tattoos among celebrity culture during the tattoo renaissance, it is not unusual that models with tattoos would become a more prominent presence in advertising campaigns. These bodies, of course, are also produced within a specific visual context that constructs a distinct discursive construction. I will return to this later in the chapter but, for example, a Guess advertisement uses the tattooed male body in order to refer to discourses of mid-twentieth century masculinity. The Guess model is shirtless, wearing Guess jeans and standing at a 3/4 profile to the camera. His bare back and upper right arm are adorned with tattoo images nostalgic of old-school tattoo design most often attributed to Sailor Jerry. Namely, a bikini-clad woman wearing high heels whose hairstyle, bikini, and shoes are all reminiscent of the mid-century. The model also has short, dark hair which is slicked back James Dean style and the patch on the jeans reads "workwear." The Guess advertisement, consequently, is a clear appropriation and reference to gender discourse in its construction tattoos as a symbol of masculinity of the "good old days."
2. *Ray Gun* (September 1999): 6–7.
3. Amy C. Sims, "Athletes Sell Skin Space to Advertisers" (April 1, 2002). Foxnews.com, http://www.foxnews.com/story/0,2933,49197,00. html (accessed December 15, 2005).
4. Associated Press, "Woman Tattoos Web Address on Forehead" (June 30, 2005). CNN.com, http://www.cnn.com/2005/US/06/30/forehead.tattoo.ap/index.html%3Fsection'cnn_offbeat (accessed December 15, 2005).
5. Butler, *Gender Trouble*.

6. John Sloop, *Disciplining Gender: Rhetorics of Sex Identity in Contemporary U.S. Culture* (Boston, MA: University of Massachusetts Press, 2004), 6. Sloop's rendition is that "gender is what we *do* rather than what we *are*" (6). The footnote at the end of this statement clarifies that Butler is not locating gender in the intentionally and material actions of a willing subject, but rather that her critique is also that the performativity of gender participates in constituting the possibility of interiority. This is why my turn of the phrase eliminates the subject "we."
7. Butler, *Gender Trouble*, 174.
8. Ibid., 6.
9. Butler, *Bodies that Matter*, 126–127.
10. As West and Zimmerman point out, however, gender articulation (or misarticulation) is not always obviously apparent. See Candace West and Don H. Zimmerman, "Doing Gender," in *The Social Construction of Gender*, ed. Judith Lorber and Susan A. Farrell (London: Sage Publications, 1991), 24–25.
11. Joan Acker, "Hierarchies, Jobs, Bodies: A Theory of Gendered Organizations," in *The Social Construction of Gender*, ed. Judith Lorber and Susan A. Farrell (London: Sage Publications, 1991), 167.
12. Perhaps the most pedestrian example of gendered social space is the public restroom and the quandary the "male" and "female" designators pose to a transgendered and/or transsexual individual. Another instance of this is the gendering of nurseries through decor (blue or pink) based on the sex of the baby.
13. Michael Kimmel, *The Gendered Society* (New York: Oxford University Press, 2000), 95.
14. *Glamour* (January 2000):101. *Glamour* is a popular women's magazine in the U.S. featuring articles about women's fashion, diet, exercise, health and social issues. According to their Web site, Glamour's U.S. circulation tops 2.3 million and the majority of readers are women between the ages of 18–39 with a median household income of $58,000. "Conde Nast Media Kit: Circulation/ Demographics" (February 2, 2005); *Glamour Home Page* (accessed January 4, 2006), http://www.condenastmediakit.com/gla/circulation.cfm.
15. There is no exact date for the poster. Melissa Dabakis, "Gendered Labor: Norman Rockwell's *Rosie the Riveter* and the Discourses of Wartime Womanhood," in *Gender in American History Since 1890*, ed. Barbara Melosh (New York: Routledge, 1993), 200.
16. Ibid., 199.
17. Ibid.
18. The one other addition/change made to the poster image is the address for the Tampax web site in small print in the bottom right-hand corner of the advertisement.
19. Roach insists, of course, that the location of an original is always fraught with difficulty because there is always an element of surrogation in any cultural text.
20. Dabakis, "Gendered Labor," 197–201.
21. Ibid., 199.

22. Ibid., 194.
23. Ibid.
24. Peiss, *Cheap Amusements*, 47. Peiss suggests in her discussion of leisure and labor at the turn of the century until 1920, for example, that "the saleslady's patina of style and refinement differentiated her from the rougher manner of many tobacco or garment workers" (47).
25. The cover of *Life Magazine* (October 1, 1923) depicted a finely made-up woman in high heels, flowing dress with lacy apron, and long curly hair pinned up riding a vacuum cleaner. The illustration was titled "A Modern Witch."
26. David R. Shumway, "Watching Elvis: The Male Rock Star as Object of the Gaze," in *The Other Fifties*, ed. Joel Foreman (Chicago, IL: University of Illinois Press, 1997), 126.
27. Dabakis, "Gendered Labor," 201.
28. Ibid., 191.
29. Ibid.
30. Sarah M. Evans, *Born For Liberty: A History of Women in America* (New York: The Free Press, 1989), 260.
31. Mitchell, *What Do Pictures Want*, 259.
32. I am referring here to, for example, the differences in assumed in the values and actions of female vs. male politicians, CEOs, soldiers, etc. While the actual job descriptions don't change, the expectations for how to fulfill those descriptions is subtly (and sometimes not so subtly) different for women than it is for men.
33. Ednie Kaeh Garrison, "U.S. Feminism—Grrrl Style! Youth (Sub)Cultures and the Technologies of the Third Wave," *Feminist Studies*, 26.1 (Spring 2000), 145.
34. Kimmel, *The Gendered Society*, 268.
35. Douglas, *Purity and Danger*, 63.
36. Ibid., 40.
37. Ibid.
38. Douglas was referring to cultural practices that actually kill anomalous humans and animals, but I think the analogy to a metaphorical death within a representational economy is compelling.
39. Govenar, "Introduction," xxii.
40. Ibid., xxiii.
41. Ibid., xxii.
42. Steward, *Bad Boys*, 58.
43. I am aware of the phallic implications of this statement.
44. Nancy Duncan, "Renegotiating Gender and Sexuality in Public and Private Spaces," in *BodySpace*, ed. Nancy Duncan (New York: Routledge, 1996), 128.
45. Bakhtin, "Chronotope," 249.
46. James Gilbert, *Another Chance: Postwar America, 1945–1985*, 2nd edition (Chicago, IL: Dorsey Press, 1986), 116.
47. Michael Kimmel, "Consuming Manhood: The Feminization of American Culture and the Recreation of the Male Body, 1832–1920," in *The Male*

Body: Features, Destinies, Exposures, ed. Laurence Goldstein (Ann Arbor, MI: University of Michigan Press, 1995), 21.

48. Gerald Early, *The Culture of Bruising: Essays on Prize Fighting, Literature, and Modern American Culture* (Hopewell, NJ: Ecco Press, 1994).
49. Early, 90.
50. Ibid., 87.
51. Steward, *Bad Boys*, 39.
52. Govenar, "Introduction," xxiii.
53. Another factor contributing to the classification of tattoos as deviant in the mid-twentieth century was an outbreak of hepatitis in the 1950s and early 1960s (see Chapter Two). As a consequence of the disease, tattooing was banned in some states. Margot Mifflin argues that these bans secured tattooing's association with deviance because they constructed tattooing as an "outlaw" practice. Within the context of the Marlboro advertisement, "outlaw" is reconfigured as freedom from social regulation. The cowboy "in some loose moment" exercised his individuality by breaking the law. Of course, the Marlboro advertisement also has the cowboy costumed in a western-style suit, implying that his youthful indiscretion has since been replaced by the culturally sanctioned practice of land-owning businessmen. The bodily freedom of expression the tattoo symbolizes is displaced to roaming the ranch and smoking "a lot" of cigarette.
54. Ted Polhemus (photographs by Housk Randall), *The Customized Body* (New York: Serpent's Tail, 1996), 24. The parallel trend for soldiers and sailors to symbolize group membership via tattooing was also becoming coded as deviant by the military as it sought to regulate the activity. For a more detailed discussion of this issue, see Chapter Two.
55. Sanders, *Customizing the Body*, 2.
56. Steward, *Bad Boys*, 63.
57. By Leonard St.Clair and Alan B. Govenar, *Stoney Knows How: Life as a Tattoo Artist* (Lexington, KY: University of Kentucky Press, 1989), 101.
58. Steward, *Bad Boys*, 57.
59. Wardell B. Pomeroy, "Foreword," in *Bad Boys and Tough Tattoos* by Samuel Steward (New York: The Haworth Press, 1990), 2.
60. Steward, *Bad Boys*, 5.
61. Bogdan, *Freak Show*, 241.
62. Sherwood, *Here We Are Again*, 149.
63. Govenar, "Introduction," xvi.
64. Kim Hewitt, *Mutilating the Body: Identity in Blood and Ink* (Bowling Green, OH: Bowling Green State University Popular Press, 1997), 75.
65. Sanders, *Customizing the Body*, 48.
66. Ibid.
67. Jake Fogelnest, "21st Century Girls," *Ray Gun* (September 1999): 25.
68. UmJammer Lammy advertisement, *Ray Gun* (September 1999): 7.
69. Laura Gilbert and Crystal Kile, *Surfergrrrrls: Look Ethel! An Internet Guide for Us!* (Seattle, WA: Seal Press, 1996). qtd in Garrison, "U.S. Feminism," 141.
70. Bogdan, *Freak Show*, 251.

71. Mifflin, *Bodies of Subversion*, 30.
72. Ibid., 32.
73. Ibid., 23.
74. Richard Bauman, "Performance," in *Folklore, Cultural Performances, and Popular Entertainments*, ed. Richard Bauman (New York: Oxford University Press, 1992), 46.
75. Steward, *Bad Boys*, 48–49.
76. Ibid., 49. Steward's experience is, of course, with a specific class of male clientele who, it seems, perceived tattooing as a masculine expression of ownership and control over their female partners. It is significant, nonetheless, because during this era tattooing was largely a lower-class practice and so cultural interpretations of tattoos and tattooing were formed within and in response to this community.
77. Margot Mifflin offers a very comprehensive descriptive analysis of women and tattoos throughout the twentieth century in the U.S. in *Bodies of Subversion*.
78. Mifflin, *Bodies of Subversion*, 39.
79. Braunberger, "Revolting Bodies," 15.
80. Tattoos on celebrities are so prevalent that MTV even noted the trend in a year-end countdown of the top nine trends of 1999.
81. Gilbert and Kile, *Surfergrrrrls*, xx.
82. Mifflin, *Bodies of Subversion*, 100, 167.
83. Braunberger, "Revolting Bodies," 15.
84. Garrison indicates in her essay that the young women of this movement recognize the ineffectiveness of mainstream media in communicating their message and so rely mainly on alternatives like the Internet, punk music, and zines. My point is, however, that the mainstream media has co-opted the image in order to efface the power of the subcultural message. Punk feminists, as a result, suffer the same fate as Hebdige suggests punk (anti)fashion endured in Britain.
85. Garrison, "U.S. Feminism," 158.
86. Jean Baudrillard, "Simulacra and Simulations," in *Selected Writings*, ed. and trans. Mark Poster (Stanford, CA: Stanford University Press, 1988), 166–184.
87. Braunberger, "Revolting Bodies," 15.
88. These potential repositionings of the advertising image activate the possibility of what Hutcheons refers to as postmodern photographic critique: "Reappropriating existing representations that are effective precisely because they are loaded with preexisting meaning and putting them in new and ironic contexts is a typical form of postmodern photographic complicitous critique: while exploiting the power of familiar images, it also denaturalizes them, makes visible the concealed mechanisms which work to make them seem transparent, and brings to the fore their politics." Linda Hutcheons, *The Politics of Postmodernism* (New York: Routledge, 1989), 44.
89. The fact that the war poster image is in the public domain has, of course, contributed to its proliferation as the image can be used without paying fees

for permission, licensing or violating copyright. What this fact suggests is that the reproductive potential of images is also implicated by the machine of capitalism.

90. A quick Google (accessed December 12, 2006) search for "Rosie the Riveter tattoo" resulted in photos of Rosie etched on the backs and arms of men and women alike.
91. Back Up Your Birth Control, http://www.backupyourbirthcontrol.org/index.htm (accessed December 12, 2006).
92. Back up Your Birth Control, "Materials Order Form," http://www.backupyourbirthcontrol.org/documents/Materials_Order_Form.pdf (accessed December 12, 2006).
93. For images of pierced hearts from the early 20th century, see Michael McCabe, *New York City Tattoo: The Oral History of an Urban Art* (Honolulu, HI: Hardy Marks Publications, 1997). For a thorough discussion of the sacred heart, see Karmen MacKendrick, *Word Made Skin* (New York: Fordham University Press, 2004), 115–136.
94. MacKendrick, *Word Made Skin*, 132.
95. Bernadette Wegenstein, "Getting Under the Skin, or, How Faces have Become Obsolete," *Configurations*, 10 (2002): 221–259.
96. Wegenstein's overarching argument is that prior to the recent power of the fragmented body part to become its own totality and overcode the unity of the human body, the face was the only part of the body that had this power.
97. The OiB is an inversion of Deleuze and Guattari's concept of the body-without-organs (BwO). Wegenstein's claim is that the utopian possibility of the democratic multiplicity of the BwO has been usurped by the overcoding of the body by individual organs. While Deleuze and Guattarri's formulation believes in the potential of the deterriorialized body, Wegenstein's perspective suggests that individual parts are themselves the surface upon which the body is produced.
98. MacKendrick, *Word Made Skin*, 142.
99. John Stratton, "Building a Better Body," in *Sportcult*, ed. Randy Martin and Toby Miller (Minneapolis, MN: University of Minnesota Press, 1999), 158.
100. Jean Luc Nancy, "Corpus," in *Thinking Bodies*, ed. Juliet Flower MacCannell and Laura Zakarin (Stanford, CA: Stanford University Press, 1994), 24.

Chapter Four Modern Primitives: Exoticism, Hybridity, and Photography

1. V. Vale and Andrea Juno, *Modern Primitives: An Investigation of Contemporary Adornment and Ritual* (San Francisco, CA: Re/Search Publications, 1989), 5.
2. DeMello, *Bodies of Inscription*, 176.
3. In particular, exoticism and visual representation have been the subjects of multiple volumes. See Peter Mason, *Infelicities: Representations of the Exotic*

(Baltimore, MD: Johns Hopkins University Press, 1998); Frederick N. Bohrer, *Orientalism and Visual Culture: Imagining Mesopotamia in Nineteenth-Century Europe* (Cambridge, UK: Cambridge University Press, 2003); Assenka Oksoloff, *Picturing the Primitive* (New York: Palgrave, 2001).

4. Harriet Guest, "Curiously Marked: Tattooing and Gender Difference in Eighteenth-century British Perceptions of the South Pacific," in *Written on the Body*, ed. Jane Caplan (Princeton, NJ: Princeton University Press, 2000), 83–101.
5. Kateryna Olijnyk Longley, "Fabricating Otherness: Demidenko and Exoticism," in "*New*" *Exoticisms*, ed. Isabel Santaolalla (Atlanta, GA: Rodopi, 2000), 23.
6. Ovidi Carbonell, "Exoticism in Translation: Writing, Representation, and the Postcolonial Context," in "*New*" *Exoticisms*, ed. Isabel Santaolalla (Atlanta, GA: Rodopi, 2000), 52.
7. For an explanation of this position see Isabel Santaolalla, "Introduction," in "*New*" *Exoticisms*, ed. Isabel Santaolalla (Atlanta, GA: Rodopi, 2000), 9.
8. Edward Said, *Orientalism* (New York: Random House, 1978), 40.
9. Craig Latrell, "After Appropriation," *The Drama Review*, 44.4 (2000): 44–55. Latrell objects to the exotic-as-appropriation logic based on its inflexibility and blindness to context (44).
10. Mason, *Infelicities*, 2.
11. Ibid.
12. Ibid., 3.
13. Bohrer, *Orientalism and Visual Culture*, 15.
14. Mason, *Infelicities*, 147–164.
15. The use of the term hybridity here is drawn from Homi Bhabha as traced within the lineage of exotic performativity described through the work of Mason and Bohrer.
16. Homi Bhabha, *The Location of Culture* (New York: Routledge, 1994), 86.
17. WWW.fakir.org
18. Vale and Juno, *Modern Primitives*, 8.
19. Modern primitive tattooing, as explained by Musafar, therefore becomes an embodied expression of Mikhail Bakhtin's concept of aesthetic seeing. Mikhail Bakhtin, "Author and Hero in Aesthetic Activity," in *Art and Answerability: Early Philosophical Essays by M.M. Bakhtin*, trans. Vadim Liapunov and Kenneth Brostrom, ed. Michael Holquist and Vadim Liapunov (Austin, TX: University of Texas Press, 1990), 4–256. Bakhtin describes the connection between the "I" and the "other" in the aesthetic act as being one of an excess of seeing. This excess is largely connected to the exterior surface of the body, according to Bakhtin, because one's own body cannot be experienced by the self as an exterior form (85). Consequently, form is created in the act of contemplation of an other and that act is inherently productive (24). Concrete dichotomies of self/other, mind/body, subjectivity/objectivity dissolve in this act as the contingency of these positions becomes produced in both process and aesthetic product; these relationships cease "to be founded on a necessary principle" (5). The aesthetic body is not only produced in its

representation, however. Its production also detaches it from any specific location or origin. The process of creating the modern primitive is indicative of an essentially human and interactive, rather than culturally specific and appropriative, condition.

20. Another performative effect of this rhetoric is a rehearsal of the social inscription vs. individual resistant expression dialectic circulating in the discourse of deviance discussed in Chapter One. Scientific discourses, you will recall, during the late nineteenth century used the rhetoric of tattoos expressing personality as a means to classify the tattooed figure as a deviant or criminal. Modern primitive discourse shifts the power relations by suggesting that instead of being signs of criminality and deviance, tattoos are expressions of humanity as well as being the product of social relations.
21. Vale and Juno, *Modern Primitives*, 8.
22. Ibid.
23. The term anchor is drawn from Roland Barthes discussion of the relation between linguistic and iconic messages. See Roland Barthes, "Rhetoric of the Image," in *Image-Music-Text*, ed. and trans. Stephen Heath (New York: Hill and Wang, 1977), 38–40.
24. This construction is similar to the effect that Rosalind Krauss describes in terms of family photography where images in photographs produce a "theater" where the family is "together and whole." Rosalind Krauss, "A Note on Photography and the Simulacral," in *Over Exposed: Essays on Contemporary Photography*, ed. Carol Squiers (New York: The New York Press, 1999), 174.
25. Mason, *Infelicities*, 160.
26. DeMello, *Bodies of Inscription*, 176.
27. Mason, *Infelicities*, 159.
28. Ibid.
29. Vale and Juno, *Modern Primitives*, 13.
30. Kobena Mercer, "Mortal Coil: Eros and Diaspora in the Photographs of Rotimi Fani-Kayode," in *Over Exposed: Essays on Contemporary Photography*, ed. Carol Squiers (New York: The New York Press, 1999), 198.
31. Vale and Juno, *Modern Primitives*, 96–100.
32. See, for example, Vale and Juno, *Modern Primitives*, 97; DeMello, *Bodies of Inscription*, 86; Victoria Lautman, *The New Tattoo* (New York: Warner Books Inc., 1994), 21.
33. Vale and Juno, *Modern Primitives*, 98.
34. Ibid., 99.
35. DeMello, *Bodies of Inscription*, 87–88.
36. Vale and Juno, *Modern Primitives*, 98. Also note that the reference to "Ed" is to Ed Hardy. Hardy is a prominent tattoo artist and also represented (although specifically not photographically) in Vale and Juno's volume.
37. Rufus Camphausen, *Return of the Tribal: A Celebration of Body Adornment* (Rochester, VT: Park Street Press, 1997), 2.
38. Ibid. Also note the association between the masculine and hand tattooing, see Chapter Three.

39. This particular photograph also appears in DeMichele's *The Illustrated Woman* (44–45). In DeMichele's text, there are five photographs of "Gloria," including the image that appears in Camphausen's text. The tattoo artist is named as Jonathan Shaw (44). Shaw is a well-known, self-proclaimed "urban" tattoo artist who operates Fun City Studio in New York City and is perhaps partly famous for tattooing celebrities such as Johnny Depp (Krakow, *Total Tattoo Book*, 100–102).
40. Chinchilla, *Stewed, Screwed and Tattooed*, 47. In a section called "Modern Day Tribal Rituals," Chinchilla notes that "bold, black, sculptural" designs that "adapt well to the natural curves of the human body" are characteristic of "new tribalism" (47).
41. Camphausen, *Return of the Tribal*, 7.
42. Polhemus, *Customized Body*, 9.
43. Ibid., 23–24.
44. Ibid., 32.
45. Ibid.
46. Bhabha, *Location of Culture*, 36.
47. Ibid.
48. Ibid., 37.
49. Ibid., 1–2.
50. Colin Turnbull, "Liminality: A Synthesis of Subjective and Objective Experience," in *By Means of Performance*, ed. Richard Schechner and Willa Appel (Cambridge, UK: Cambridge University Press, 1990):50–81. Turnbull argues that transformation, as opposed to transition, is the defining element of liminal experience (79).
51. Bhabha, *Location of Culture*, 39.
52. Camphausen, *Return of the Tribal*, 10.
53. Polhemus, *Customized Body*, 24.
54. Vale and Juno, *Modern Primitives*, 37.
55. Ibid.
56. Vale and Juno, *Modern Primitives*, 39.
57. Ibid., 38.
58. As a matter of form, Vale and Juno's questions are in italics, which forges a link between italicized text in the interviews and the interviewer's voice.
59. Val and Juno, *Modern Primitives*, 39.
60. Ibid., 5.
61. Ibid.
62. Ibid., 39.
63. Ibid.
64. Ibid. The term "psychedelia" also alludes, of course, to the practice of using drugs to transport the mind; to have a disembodied experience. It is interesting that drugs are also a focus in *Return of the Tribal*. The mind/body dichotomy disintegrates yet again because the practice of painful body alteration is compared to the effects of psychedelic drugs; spiritual/mental transformation occurs through embodied practices.
65. Vale and Juno, *Modern Primitives*, 68.

66. Ibid., 68–75.
67. Ibid., 75.
68. Vale and Juno, *Modern Primitives*, 70–71.
69. Captain Don also indicates that Fakir Musafar used to be a carnival showman before "moving on to other things" (Vale and Juno, *Modern Primitives*, 73).
70. Vale and Juno, *Modern Primitives*, 68.
71. Ibid., 73.
72. Bogdan explains that tattoo exhibitions waned in the 1930's and "some tattooed men resorted to bizarre performances to make themselves more marketable on the highly competitive freak market" (256).
73. Vale and Juno, *Modern Primitives*, 73.
74. Ibid., 74.
75. Ibid.
76. Ibid., 75.
77. Ibid.
78. Ibid.
79. Ibid.
80. Bogdan, *Freak Show*, 105.
81. Ibid., 241.
82. Ibid., 242.
83. George Burchett, *Memoirs of a Tattooist* (New York: Crown, 1958), 24. Rutherford's experience also set the precedent for seamen who, upon returning from the sea and/or military service, found a career in exhibiting their tattooed bodies. See Bogdan, *Freak Show*, 241. The connection between American naval servicemen and tattoos can also be traced to this tradition.
84. Sherwood, *Here We Are Again*, 148–150.
85. See S. Gilbert, *Tattoo History*, 137.
86. Ibid.
87. Mifflin, *Bodies of Subversion*, 10.
88. Cesare Lombroso, "The Savage Origin of Tattooing" *Popular Science Monthly* (April 1896): 800.
89. Bogdan, *Freak Show*, 249.
90. Ibid., 105–107.
91. Ibid., 105.
92. *New York Times* qtd. in Mifflin, *Bodies of Subversion*, 10.
93. Sherwood, *Here We Are Again*, 151.
94. Bogdan, *Freak Show*, 249.
95. Polhemus, *Customized Body*, 30.
96. Ibid., 36.
97. Bogdan, *Freak Show*, 106.
98. Ibid., 247.
99. DeMello, *Bodies of Inscription*, 145.
100. Vale and Juno, *Modern Primitives*, 5.

101. Camphausen, *Return of the Tribal*, 11. The time-line takes up pages 5–13. The history also jumps from 1891 to 1939, making only a brief reference to the performance of illustrated exhibits at fairgrounds in 1882. Sideshow performances are thus also folded into the overarching narrative of tribalism.
102. Camphausen, *Return of the Tribal*, 11.
103. Ibid.
104. Polhemus, *Customized Body*, 24.
105. Ibid.
106. DeMello, *Bodies of Inscription*, 174. Because *Modern Primitives* marks this transition, it may be the case that, unlike its followers, there was more need to account for popular working-class definitions and representations of body modification within this new narrative.
107. Vale and Juno, *Modern Primitives*, 61.
108. Ibid., 60. Interestingly, Hardy's tattooed body is not, however, pictured in *Modern Primitives*. Perhaps the aesthetic of Tattoo Mike (facial tattoos) and the historic fetishizing of Captain Don (as a "survivor of the forgotten era") more easily fit into Vale and Juno's narrative than does Hardy's graffitied body.
109. Christopher Innes, *Avant Garde Theatre, 1892–1992* (New York: Routledge, 1993), 18, 192.
110. Frederic Jameson, "Postmodernism, or the Cultural Logic of Late Capitalism," *New Left Review*, 146 (July–August 1984): 65–66.
111. Ibid., 66.
112. Ibid., 65.
113. Hal Foster, *Recodings: Art, Spectacle, Cultural Politics* (Port Townsend, WA: Bay Press, 1985), 196.
114. Gilles Deleuze, *Foucault*, ed. and trans. Sean Hand (Minneapolis, MN: University of Minnesota Press, 1988); Gilles Deleuze, *The Fold*.
115. Conley, "From Multiplicities to Folds," 640.
116. Deleuze, *Foucault*, 109.
117. Ibid.
118. The agency of the tattooed body as image here is different from (although related to) the space of resistance that is provided when those who don't identify with middle-class and modern primitivist discourse speak out against that discourse. In the section of her study "Tattoo Magazines II: Bikers Write Back," DeMello suggests that certain "low brow" (DeMello's term) tattoo magazines have become a site for members of the tattoo community excluded by modern primitivist-like rhetoric and performance to express their disaffiliation (113–124). DeMello includes excerpts from reader's letters published in magazines like *Tattoo*, which began as a biker magazine. In one particular series, letter writers respond to a previously published letter whose author suggested that biker-style tattoos were ugly and project a negative image. The biker respondents were thus given a discursive space to articulate their disidentified identity. In one particularly colorful response to the critique, a writer says: "So you think they are ugly and project an image that turns off many people. All I can say is, 'Fuck you

and your long fingernails' "(115). This response and others like it indicate that tattoo enthusiasts who do not adhere to the modern primitive philosophy tend to view those who do as elitist and exclusionary. They mock the distinction made between tattoos that are old-school or trivial and those that represent neotribal belonging and thus they highlight the artificiality of the difference.

Conclusion: Imag(e)Ining Agency and Criticism

1. Rebecca Schneider, "Intermediality, Infelicity, and Scholarship on the Slip," *Theatre Survey* 47 (November 2006): 253–260.
2. This is not to say that the process and product of criticism in general, and the role and value of the interpretation of meaning in particular, has not been long a topic of discussion. See, for example, Susan Sontag, *Against Interpretation* (New York: Farrar, Straus & Giroux, 1966).
3. Gavin Butt, "Introduction: The Paradoxes of Criticism," in *After Criticism: New Responses to Art and Performance*, ed. Gavin Butt (Malden, MA: Blackwell, 2005), 1.
4. Gumbrecht, *Production of Presence*, 1.
5. Matthew Reason, *Documentation, Disappearance and the Representation of Live Performance* (New York: Palgrave Macmillan, 2006).
6. Della Pollock, "Performing Writing," in *The Ends of Performance*, ed. Peggy Phelan and Jill Lane (New York: New York University Press, 1998), 73–103.
7. Gerry Brenner, *Performative Criticism: Experiments in Reader Response* (Albany, NY: State University of New York Press, 2004).
8. Craig Gingrich-Philbrook, "Autoethnography's Family Values: Easy Access to Compulsory Experiences," *Text and Performance Quarterly*, 25 (2005): 299.
9. Although, as Gingrich-Philbrook aptly points out, the attention to style and aesthetics in some of the discourse lacks precision, sophistication, and answerability to the history, philosophy and practices of aesthetics and poetics.
10. Tami Spry, "Preface to the Proceedings," in *The Green Window: Proceedings of the Giant City Conference on Performative Writing*, ed. Lynn C. Miller and Ronald J. Pelias (Carbondale, IL: Southern Illinois University, 2001), viii–ix.
11. Jonathan Gray, "Preface to the Proceedings," in *The Green Window: Proceedings of the Giant City Conference on Performative Writing*, ed. Lynn C. Miller and Ronald J. Pelias (Carbondale, IL: Southern Illinois University, 2001), vii–viii.
12. Some styles are compilation of textual fragments, poetry, personal narrative, photo essays, autoethnography, etc. For some excellent examples of performative writing that is conceptually, critically, and creatively engaged, see: Michael Bowman, "Killing Dillinger," *Text and Performance Quarterly*, 20 (2000): 342–375; Ruth Laurion Bowman and Michael S. Bowman, "On The Bias: From Performance of Literature to Performance Composition," in *The Sage Handbook of Performance Studies*, ed. D. Soyini Madison and Judith Hamera (Thousand Oaks, CA: Sage, 2006): 205–226.

13. Ronald J. Pelias, *A Methodology of the Heart: Evoking Academic and Daily Life* (Walnut Creek, CA: Alta Mira Press, 2004).
14. I have explored the operation of these dialectic oppositions in terms of performance practice and criticism at length elsewhere. See Mindy Fenske, "The Aesthetic of the Unfinished: Ethics and Performance," *Text and Performance Quarterly*, 24 (2004): 1–19; Mindy Fenske, "The Movement of Interpretation: Conceptualizing Performative Encounters with Multimediated Performance," *Text and Performance Quarterly*, 26 (2006):138–161.
15. Gingrich-Philbrook, "Autoethnography's Family Values," 306.
16. Butt, "Introduction," 3–5.
17. Ibid., 4.
18. The same criticism has been leveled at many authors who revel in the complexities and nuances of language—authors who expect much of their readers. Judith Butler, for example, has suffered from this attack.
19. See Frederick C. Corey, "Alexander," *Communication Quarterly*, 50 (2002): 344–358; Pollock, "Performing Writing."
20. Mary Poovey, "Creative Criticism: Adaptation, Performative Writing, and the Problem of Objectivity," *Narrative*, 8.2 (2000), 123.
21. Elsewhere I have prompted performative criticism based on the models of the virus (infection) and termite (consumption). See Fenske, "The Movement of Interpretation."

BIBLIOGRAPHY

Acker, Joan. "Hierarchies, Jobs, Bodies: A Theory of Gendered Organizations." In *The Social Construction of Gender*, ed. Judith Lorber and Susan A. Farrell, 162–179. London: Sage Publications, 1991.

Allison, Dorothy. *Skin: Talking About Sex, Class, and Literature*. Ithaca, NY: Firebrand, 1994.

Anderson, Clare. "*Godna*: Inscribing Indian Convicts in the Nineteenth Century." In *Written onthe Body*, ed. Jane Caplan, 102–117. Princeton, NJ: Princeton University Press, 2000.

Armstrong, Myrna L., D. J. Stuppy, D. C. Gabriel, and R. R. Anderson "Motivation for Tattoo Removal." *Archives of Dermatology* 132 (1996): 412–416.

Armstrong, Tim. "Introduction to *American Bodies*." In *American Bodies*, ed. Tim Armstrong, 9–17. New York: New York University Press, 1996.

Associated Press. "Woman Tattoos Web Address on Forehead." June 30, 2005. CNN.com.http://www.cnn.com/2005/US/06/30/forehead.tattoo.ap/index.html%3Fsection'cnn_offbeat (accessed December 15, 2005).

Austin, J.L. *How to Do Things With Words*. Cambridge, MA: Harvard University Press, 1975.

Bakhtin, M.M. "Author and Hero in Aesthetic Activity." In *Art and Answerability: Early Philosophical Essays by M.M. Bakhtin*, ed. Michael Holquist and Vadim Liapunov, trans. Vadim Liapunov and Kenneth Brostrom, 4–256. Austin, TX: University of Texas Press, 1990.

———. "Forms of Time and Chronotope in the Novel." In *The Dialogic Imagination*, ed. Michael Holquist, trans. Caryl Emerson, 84–258. Austin, TX: University of Texas Press, 1981.

———. *Toward a Philosophy of the Act*, ed. Vadim Liapunov and Michael Holquist, trans. Vadim Liapunov. Austin, TX: University of Texas Press, 1993.

Barthes, Roland. *Camera Lucida*. New York: Hill and Wang, 1980.

———. *Image-Music-Text*. New York: Hill and Wang, 1977.

———. *Mythologies*. New York: Hill and Wang, 1994.

Baudrillard, Jean. "Simulacra and Simulations." In *Selected Writings*, ed. and trans. Mark Poster, 166–184. Stanford, CA: Stanford University Press, 1988.

Bauman, Richard. "Performance." In *Folklore, Cultural Performances, and Popular Entertainments*, ed. Richard Bauman, 41–49. New York: Oxford University Press, 1992.

Benjamin, Walter. "The Work of Art in the Age of Mechanical Reproduction." In *Illuminations*, ed. Hannah Arendt, trans. Harry Zohn. New York: Random House, 1988.

Benson, Susan. "Inscriptions of the Self: Reflections on Tattooing and Piercing in Contemporary Euro-America." In *Written on the Body*, ed. Jane Caplan, 234–254. Princeton, NJ: Princeton University Press, 2000.

Bhabha, Homi. *The Location of Culture*. New York: Routledge, 1994.

Bianchi, Robert S. "Tattoo in Ancient Egypt." In *Marks of Civilization: Artistic Transformations of the Human Body*, ed. Arnold Rubin, 21–29. Los Angeles, CA: Museum of Cultural History, University of California, 1988.

Blocker, Jane. *What the Body Cost: Desire, History, and Performance*. Minneapolis, MN: University of Minnesota Press, 2004.

Bogdan, Robert. *Freak Show*. Chicago, IL: University of Chicago Press, 1988.

Bohrer, Frederick N. *Orientalism and Visual Culture: Imagining Mesopotamia in Nineteenth-Century Europe*. Cambridge, UK: Cambridge University Press, 2003.

Bourdieu, Pierre. "Distinction: A Social Critique of the Judgement of Taste." In *Social Stratification: Class, Race, and Gender in Sociological Perspective*, ed. David B. Grusky, 404–429. Boulder, CO: Westview Press, 1994.

Bowman, Michael S. "Killing Dillinger." *Text and Performance Quarterly* 20 (2000): 342–375.

Bowman, Ruth Laurion, and Michael S. Bowman. "On the Bias: From Performance of Literature to Performance Composition." In *The Sage Handbook of Performance Studies*, ed. D. Soyini Madison and Judith Hamera, 205–226. Thousand Oaks, CA: Sage, 2006.

Bradley, James. "Body Commodification? Class and Tattoos in Victorian Britain." In *Written on the Body*, ed. Jane Caplan, 136–155. Princeton, NJ: Princeton University Press, 2000.

Brain, Robert. *The Decorated Body*. New York: Harper and Row Publishers, 1979.

Braunberger, Christine. "Revolting Bodies: The Monster Beauty of Tattooed Women." *NWSA Journal* 12, no. 2 (2000): 1–23.

Brecht, Bertold. *Brecht on Theatre*, ed. and trans. John Willet. New York: Hill and Wang, 1992.

Brenner, Gerry. *Performative Criticism: Experiments in Reader Response*. Albany, NY: State University of New York Press, 2004.

Briggs, J. "Tattooing." *Medical Times* 87 (1958): 1030–1039.

Brouwer, Dan. "The Precarious Visibility Politics of Self-Stigmatization: The Case of HIV/AIDS Tattoos." *Text and Performance Quarterly* 18 (1998): 114–136.

Burchett, George. *Memoir of a Tattooist*. New York: Crown, 1958.

Butler, Judith. *Bodies That Matter*. New York: Routledge, 1993.

———. *Gender Trouble*. New York: Routledge, 1990.

Butt, Gavin. "Introduction: The Paradoxes of Criticism." In *After Criticism: New Responses to Art and Performance*, ed. Gavin Butt, 1–20. Malden, MA: Blackwell Publishing, 2005.

Camphausen, Rufus C. *Return of the Tribal*. Rochester, VT: Park Street Press, 1997.

Caplan, Jane. "'National Tattooing': Traditions of Tattooing in Nineteenth-century Europe." In *Written on the Body*, ed. Jane Caplan, 156–173. Princeton, NJ: Princeton University Press, 2000.

Carbonell, Ovidi. "Exoticism in Translation: Writing, Representation, and the Postcolonial Context." In *"New" Exoticisms*, ed. Isabel Santaolalla, 511–564. Atlanta, GA: Rodopi, 2000.

Carlson, Marvin. *Performance: A Critical Introduction*. New York: Routledge, 1996.

Cassuto, Leonard. "'What an Object You Have Made of Me!' Tattooing and the Racial Freak in Melville's *Typee*." In *Freakery: Cultural Spectacles of the Extraordinary Body*, ed. Rosemarie Garland Thomson, 234–247. New York: New York University Press, 1996.

Chinchilla, Madame. *Stewed, Screwed and Tattooed*. Fort Bragg, CA: Isadore Press, 1997.

"Conde Nast Media Kit: Circulation/Demographics." February 2, 2005 *Glamour Home Page* January 4, 2006, http://www.condenastmediakit.com/gla/ circulation.cfm.

Conley, Tom. "From Multiplicities to Folds: On Style and Form in Deleuze." *The South Atlantic Quarterly* 96 (Summer 1997): 629–646.

Corey, Frederick C. "Alexander." *Communication Quarterly* 50 (2002): 344–358.

Crary, Jonthan. *Techniques of the Observer: On Vision and Modernity in the 19th Century*. Cambridge, MA: MIT Press, 1992.

Dabakis, Melissa. "Gendered Labor: Norman Rockwell's *Rosie the Riveter* and the Discourses of Wartime Womanhood." In *Gender in American History Since 1890*, ed. Barbara Melosh, 182–204. New York: Routledge, 1993.

de Certeau, Michel. *The Practice of Everyday Life*. Berkeley, CA: University of California Press, 1984.

Deleuze, Gilles. *Difference and Repetition*, trans. Paul Patton. New York: Columbia University Press, 1994.

———. *The Fold: Leibniz and the Baroque*, trans. Tom Conley. Minneapolis, MN: University of Minnesota Press, 1993.

———. *Foucault*, trans. Sean Hand. Minneapolis, MN: University of Minnesota Press, 1986.

Deleuze, Gilles, and Felix Guattari. *A Thousand Plateaus*, tans. Brian Massumi. Minneapolis, MN: University of Minnesota Press, 1987.

Deluca, Kevin. *Image Politics*. New York: The Guilford Press, 1999.

DeMello, Margo. *Bodies of Inscription: A Cultural History of the Modern TattooCommunity*. Durham, NC: Duke University Press, 2000.

———. "The Carnivalesque Body: Women and Tattoos." In *Pierced Hearts and True Love: A Century of Drawings for Tattoos*, ed. Don Ed Hardy, 72–79. Honolulu, HI: Hardy Marks Publications/New York: The Drawing Center, 1995.

DeMichele, William. *The Illustrated Woman*. New York: Proteus Press, 1992.

Dennet, Andrea Stulman. "The Dime Museum Freak Show Reconfigured as Talk Show." In *Freakery: Cultural Spectacles of the Extraordinary Body*, ed. Rosemarie Garland Thomson, 315–326. New York: New York University Press, 1996.

Derrida, Jacques. "Limited Inc abc." In *Limited Inc.*, ed. Gerald Graff, 29–110. Evanston, IL: Northwestern University Press, 1988.

———. *Of Grammatology*. London: John Hopkins Press, 1976.

Derrida, Jacques. "Signature Event Context." In *Limited Inc*, ed. Gerald Graff, 1–23. Evanston, IL: Northwestern University Press, 1988.

DeVere Brody, Jennifer. "Black Cat Fever: Manifestations of Manet's *Olympia*." *Theatre Journal* 53 (2001): 95–118.

Doenecke, Justus D. "Myths, Machines and Markets: The Columbian Exposition of 1893." *Journal of Popular Culture* 7(1973): 535–549.

Douglas, Mary. *Purity and Danger: An Analysis of Concepts of Pollution and Taboo*. Harmondsworth, UK: Penguin Books, 1970.

Dreyfus, Hubert L., and Rabinow, Paul. *Michel Foucault: Beyond Structuralism and Hermenuetics*. Chicago, IL: The University of Chicago Press, 1983.

DuGay, Paul. *Production of Culture/Cultures of Production*. Culture Media and Identities 4. London: Sage, 1997.

Duncan, Nancy. Introduction to *BodySpace*, ed. Nancy Duncan, 211–218. Seattle, WA: Seal Press, 2000.

Early, Gerald. "The Romance of Toughness." In *The Culture of Bruising: Essays on Prize Fighting, Literature, and Modern American Culture*, ed. Gerald Early, 86–109. Hopewell, NJ: Ecco Press, 1994.

Easto, Patrick C. and Marcello Truzzi. "The Carnival Social System." *Journal of Popular Culture* 7 (1973): 550–566.

Ehrenreich, Barbara. *Fear of Falling: The Inner Life of the Middle Class*. New York: Harper Perennial, 1989.

Evans, Sarah M. *Born for Liberty: A History of Women in America*. New York: The Free Press, 1989.

Falk, Robert P., and Frances Teague. "Parody." In *The New Princeton Handbook of Poetic Terms*, ed. T.V.F. Brogan, 219–221. Princeton, NJ: Princeton University Press, 1994.

Featherstone, Michael. "The Body in Consumer Culture." In *The Body: Social Process and Cultural Theory*, ed. Mike Featherstone, Mike Hepworth, and Bryan S. Turner, 170–193. London: Sage, 1991.

Fenske, Mindy. "The Aesthetic of the Unfinished: Ethics and Performance." *Text and Performance Quarterly* 24 (2004): 1–19.

———. "The Movement of Interpretation: Conceptualizing Performative Encounters with Multimediated Performance." *Text and Performance Quarterly* 26 (2006): 138–161.

Fiedler, Leslie. *Freaks: Myths and Images of the Secret Self*. New York: Simon and Schuster, 1978.

Fiske, John. *Reading the Popular*. Boston, MA: Unwin Hyman, 1989.

Fleming, Juliet. "The Renaissance Tattoo." In *Written on the Body*, ed. Jane Caplan, 61–82. Princeton, NJ: Princeton University Press, 2000.

Fogelnest, Jake. "21st Century Girls." *Ray Gun* September 1999: 25.

Foreman, Joel. Introduction to *The Other Fifties*, ed. Joel Foreman, 1–23. Chicago, IL: University of Illinois Press, 1997.

Foster, Hal. *Recodings: Art, Spectacle, Cultural Politics*. Port Townsend, WA: Bay Press, 1985.

Foucault, Michel. *Discipline and Punish*, trans. Alan Sheridan. New York: Vintage Books, 1979.

———. "Nietzche, Genealogy, History." In *Language, Counter-Memory, Practice*, ed. Donald F. Bouchard, 139–164. Ithaca, NY: Cornell University Press, 1977.

Frank, Arthur W. "For a Sociology of the Body: An Analytical Review." In *The Body*, ed. Mike Featherstone, Mike Hepworth, and Bryan Turner, 36–102. London: Sage, 1991.

Freeland, David. "Last Dance at the Orpheum." *New York Press* 18, no. 32 (2005), http://www.nypress.com/18/32/news&columns/DavidFreeland.cfm.

Fried, Fred, and Mary Fried. *America's Forgotten Folk Arts*. New York: Pantheon Books, 1978.

Garrison, Ednie Kaeh. "U.S. Feminism—Grrrl Style! Youth (Sub)Cultures and the Technologies of the Third Wave." *Feminist Studies* 26 (Spring 2000): 141–170.

Gathercole, Peter. "Contexts of Maori *Moko*." In *Marks of Civilization: Artistic Transformations of the Human Body*, ed. Arnold Rubin, 171–179. Los Angeles, CA: Museum of Cultural History, University of California, 1988.

Gerber, David A. "The Careers of People Exhibited in Freak Shows: The Problem of Volition and Valorization." In *Freakery: Cultural Spectacles of the Extraordinary Body*, ed. Rosemarie Garland Thomson, 38–54. New York: New York University Press, 1996.

Gilbert, James. *Another Chance: Postwar America, 1945–1985*. 2nd edition. Chicago, IL: Dorsey Press, 1986.

Gilbert, Steve. *Tattoo History: A Source Book*. New York: Juno Books, 2000.

Gingrich-Philbrook, Craig. "Autoethnography's Family Values: Easy Access to Compulsory Experiences." *Text and Performance Quarterly* 25 (2005): 297–314.

Goffman, Erving. *Presentation of Self in Everyday Life*. New York: Anchor, 1959.

Goldstein, Norman. "Psychological Implications of Tattoos." *Journal of Dermatologic Surgery and Oncology* 5 (1979): 883–888.

Gould, Henry. "The Unhappy Performative." In *Performativity and Performance*, ed. Andrew Parker and Eve Kosofsky Sedgwick, 19–44. New York: Routledge, 1995.

Govenar, Alan B. "The Changing Image of Tattooing in American Culture, 1846–1966." In *Written on the Body*, ed. Jane Caplan, 212–233. Princeton, NJ: Princeton University Press, 2000.

———. "Continuity and Change in the Aesthetics of Tattooing." In *Pierced Hearts and True Love: A Century of Drawings for Tattoos*, ed. Don Ed Hardy, 80–87. Honolulu, HI: Hardy Marks Publications/New York: The Drawing Center, 1995.

———. "Introduction." In *Stoney Knows How: Life as a Tattoo Artist*, by Leonard St. Clair and Alan B. Govenar, xi–xxxi. Lexington: University of Kentucky Press, 1981.

Gray, Christopher. "Streetscapes/Roseland: An Old-Fashioned Dance to the Music of Time." *New York Times* October 13, 1996. late edition, sec. 9: 5.

Gray, Jonathan. "Preface to the Proceedings." In *The Green Window: Proceedings of the Giant City Conference on Performative Writing*, ed. Lynn C. Miller and Ronald J. Pelias, vii–viii. Carbondale, IL: Southern Illinois University, 2001.

Gritton, Joy. "Labrets and Tattooing in Native Alaska." In *Marks of Civilization: Artistic Transformations of the Human Body*, ed. Arnold Rubin, 181–191. Los Angeles, CA: Museum of Cultural History, University of California, 1988.

Grossberg, Lawrence. "The Victory of Culture: Part One." *Angelaki: Journal of Theoretical Humanities* 3, no. 3 (1998): 3–30.

Grosz, Elizabeth. *Volatile Bodies: Toward a Corporeal Feminism.* Bloomington, IN: Indiana University Press, 1994.

Guest, Harriet. "Curiously Marked: Tattooing and Gender Difference in Eighteenth-century British Perceptions of the South Pacific." In *Written on the Body*, ed. Jane Caplan, 83–101. Princeton, NJ: Princeton University Press, 2000.

Gumbrecht, Hans Ulrich. *Production of Presence: What Meaning Cannot Convey.* Stanford, CA: Stanford University Press, 2004.

Gustafson, Mark. "The Tattoo in the Later Roman Empire and Beyond." In *Written on the Body*, ed. Jane Caplan, 17–32. Princeton, NJ: Princeton University Press, 2000.

Hall, Stuart. "Cultural Studies and its Theoretical Legacies." In *The Cultural Studies Reader*. 2nd edition, ed. Simon During, 97–112. New York: Routledge, 1999.

———. "Encoding, Decoding." In *The Cultural Studies Reader*. 2nd edition, ed. Simon During, 507–517. New York: Routledge, 1999.

Hambly, W.D. *The History of Tattooing and Its Significance.* Detroit: Gale Research, 1974.

Hammer, Carl, and Gideon Basker. *Freak Show: Sideshow Banner Art.* San Francisco, CA: Chronicle Books, 1996.

Hardy, Don Ed. "Current Events." In *Tattoo History: A Source Book*, ed. Steve Gilbert, 196–207. New York: Juno Books, 2000.

———. "Tattooing as a Medium." In *Pierced Hearts and True Love: A Century of Drawings for Tattoos*, ed. Don Ed Hardy, 14–27. Honolulu, HI: Hardy Marks Publications/New York: The Drawing Center, 1995.

Hariman, Robert, and John Lucaites. "Performing Civic Identity: The Iconic Photograph of the Flag Raising on Iwo Jima." *Quarterly Journal of Speech* 88 (2002): 363–393.

Harris, Neil. *Cultural Excursions.* Chicago: University of Chicago Press, 1990.

Haynes, Deborah J. *Bakhtin and the Visual Arts.* New York: Cambridge University Press, 1995.

Hebdige, Dick. *Subculture: The Meaning of Style.* London: Methuen, 1979.

Hewitt, Kim. *Mutilating the Body: Identity in Blood and Ink.* Bowling Green, OH: Bowling Green State University Popular Press, 1997.

Hewitt, Martin. "Bio-Politics and Social Policy: Foucault's Account of Welfare." In *The Body: Social Process and Cultural Theory*, ed. Mike Featherstone, Mike Hepworth, and Bryan S. Turner, 225–255. London: Sage, 1991.

Hoesterey, Ingeborg. *Pastiche: Cultural Memory in Art, Film, Literature.* Bloomington, IN: Indiana University Press, 2001.

Horrocks, Roger. *Male Myths and Icons: Masculinity in Popular Culture.* New York: St. Martin's Press, 1995.

Hutcheon, Linda. *The Politics of Postmodernism.* New York: Routledge, 1989.

Innes, Christopher. *Avant Garde Theatre, 1892–1992.* New York: Routledge. 1993.

Irwin, Katherine. "Saints and Sinners: Elite Tattoo Collectors and Tattooists as Positive and Negative Deviants." *Sociological Spectrum* 23 (2003): 27–57.

Jackson, Peter. *Maps of Meaning.* New York: Routledge, 1992.

Jameson, Fredric. "Postmodernism, or the Cultural Logic of Late Capitalism." *New Left Review* 146 (July–August 1984): 53–92.

Jones, Amelia. *Body Art: Performing the Subject.* Minneapolis, MN: University of Minnesota Press, 1998.

Jones, C.P. "Stigma and Tattoo." In *Written on the Body*, ed. Jane Caplan, 1–16. Princeton, NJ: Princeton University Press, 2000.

Kellner, Douglas. "Reading Images Critically: Toward a Postmodern Pedagogy." *Journal of Education* 170, no. 3 (1988): 31–52.

Kimmel, Michael. "Consuming Manhood: The Feminization of American Culture and the Recreation of the Male Body, 1832–1920." In *The Male Body: Features, Destinies, Exposures*, ed. Laurence Goldstein, 12–41. Ann Arbor, MI: University of Michigan Press, 1995.

———. *The Gendered Society.* New York: Oxford University Press, 2000.

Kirshenblatt-Gimblett, Barbara. *Destination Culture: Tourism, Museums, and Heritage.* Berkley, CA: University of California Press, 1998.

Korn, Kenneth. "Body Adornment and Tattooing: Clinical Issues and State Regulations." *Physician Assistant* 20, no. 5 (1996): 85.

Kosut, Mary. "Mad Artists and Tattooed Perverts: Deviant Discourse and the Social Construction of Cultural Categories." *Deviant Behavior* 27 (2006): 73–95.

Krakow, Amy. *The Total Tattoo Book.* New York: Warner Books, Inc., 1994.

Krauss, Rosalind. "A Note on Photography and the Simulacral." In *Overexposed: Essays on Contemporary Photography*, ed. Carol Squiers, 15–27. New York: The New Press, 1999.

Langellier, Kristin. "'You're Marked': Breast Cancer, Tattoo, and the Narrative Performance of Identity." In *Narrative and Identity: Studies in Autobiography, Self, and Culture*, ed. Jens Brockmeier and Doval Carbaugh, 145–184. Amsterdam and Philadelphia: John Benjamins, 2001.

Lash, Scott. "Geneaology and the Body: Foucault/Deleuze/Nietzsche." In *The Body: Social Process and Cultural Theory*, ed. Mike Featherstone, Mike Hepworth, and Bryan S. Turner, 256–280. London: Sage, 1991.

Latrell, Craig. "After Appropriation." *The Drama Review* 44, no. 4 (2000): 44–55.

Lautman, Victoria. *The New Tattoo.* New York: Abbeville Press, 1994.

Librach, Phyllis Brasch. "Much Ado About Tattoos." *St. Louis Press Dispatch* 11 (May 1997), Five Star Lift edition: C1.

"Life Goes to a Party." *Life* 2, 15 (April 12, 1937): 82–85.

Lippert, Paul. "Tattooing: The Image Becomes the Self." *ETC* 37, 2 (1980): 158–171.

Lombroso, Cesare. "The Savage Origin of Tattooing." *Popular Science Monthly* (April 1896): 793–803.

Lombroso, Cesare, and William Ferrero. *Female Offender.* New York: Philosophical Library, 1958.

Longley, Kateryna Olijnyk. "Fabricating Otherness: Demidenko and Exoticism." In "*New*" *Exoticisms*, ed. Isabel Santaolalla, 21–40. Atlanta, GA: Rodopi, 2000.

MacKendrick, Karmen. *Word Made Skin.* New York: Fordham University Press, 2004.

MacQuarrie, Charles W. "Insular Celtic Tattooing: History, Myth and Metaphor." In *Written on the Body*, ed. Jane Caplan, 32–45. Princeton, NJ: Princeton University Press, 2000.

Mascia-Lees, Frances E., and Sharpe, Patricia. "Introduction: Soft-Tissue Modification and the Horror Within." In *Tattoo, Torture, Mutilation, and Adornment*, ed. Frances E. Mascia-Lees and Patricia Sharpe, 1–9. New York: State University of New York Press, 1992.

———. "The Marked and the Un(re)marked: Tattoo and Gender in Theory and Narrative." In *Tattoo, Torture, Mutilation, and Adornment*, ed. Frances E. Mascia-Lees and Patricia Sharpe, 145–170. New York: State University of New York Press, 1992.

Mason, Peter. *Infelicities: Representations of the Exotic*. London: The Johns Hopkins University Press, 1998.

Massumi, Brian. *Parables for the Virtual*. Durham, NC: Duke University Press, 2002.

Matza, David. *Becoming Deviant*. Englewood Cliffs, NJ: Prentice-Hall, Inc., 1969.

Maxwell-Stewart, Hamish, and Ian Duffield. "Skin Deep Devotions: Religious Tattoos and Convict Transportation in Australia." In *Written on the Body*, ed. Jane Caplan, 118–135. Princeton, NJ: Princeton University Press, 2000. McCabe, Michael. *New York City Tattoo: The Oral History of an Urban Art*. Honolulu: Hardy Marks Publications, 1997.

McLaren, Peter. "Schooling the Postmodern Body: Critical Pedagogy and the Politics of Enfleshment." *Journal of Education* 170, no. 3 (1988): 53–78.

Melosh, Barbara. "Introduction." In *Gender and American History Since 1890*, ed. Barbara Melosh, 1–17. New York: Routledge, 1993.

———. "Manly Work: Public Art and Masculinity in Depression America." In *Gender and American History Since 1890*, ed. Barbara Melosh, 155–181. New York: Routledge, 1993.

Mercer, Kobena. "Mortal Coil: Eros and Diaspora in the Photographs of Rotimi Fani-Kayode" In *Over Exposed: Essays on Contemporary Photography*, ed. Carol Squiers, 183–210. New York: The New Press, 1999.

Mifflin, Margot. *Bodies of Subversion: A Secret History of Women and Tattoo*. New York: Juno Books, 1997.

Mitchell, W.J.T. *Picture Theory*. Chicago: University of Chicago Press, 1994.

———. *What Do Pictures Want*. Chicago: University of Chicago Press, 2005.

Morson, Gary Saul, and Caryl Emerson. *Mikhail Bakhtin: Creation of a Prosaics*. Stanford, CA: Stanford University Press, 1990.

Nancy, Jean Luc. "Corpus." In *Thinking Bodies*, ed. Juliet Flower MacCannell and Laura Zakarin, 17–31. Stanford, CA: Stanford University Press, 1994.

Newitz, Annalee and Matthew Wray. "What is 'White Trash'? Stereotypes and Economic Conditions of Poor Whites in the United States." In *Whiteness: A Critical Reader*, ed. Mike Hill, 168–184. New York: New York University Press, 1997.

Newman, Gustav. "The Implications of Tattooing in Prisoners." *Journal of Clinical Psychology* 43 (1982): 231–34.

Newman, Simon P. "Wearing Their Hearts on Their Sleeves: Reading the Tattoos of Early American Seafarers." In *American Bodies*, ed. Tim Armstrong, 18–31. New York: New York University Press, 1996.

Odell, George. *Annals of the New York Stage*. Vol. 4. New York: Columbia University Press, 1928.

Oettermann, Stephan. "On Display: Tattooed Entertainers in America and Germany." In *Written on the Body*, ed. Jane Caplan, 193–211. Princeton, NJ: Princeton University Press, 2000.

Oksoloff, Assenka. *Picturing the Primitive*. New York: Palgrave, 2001.

Ott, Brian. "(Re)Locating Pleasure in Media Studies: Toward an Erotics of Reading," *Communication and Critical/Cultural Studies* 1 (2004): 194–212.

Peiss, Kathy. *Cheap Amusements*. Philadelphia, PA: Temple University Press, 1986.

Pelias, Ronald J. *A Methodology of the Heart: Evoking Academic and Daily Life*. Walnut Creek, CA: Alta Mira Press, 2004.

Perry, Elisabeth. "Cleaning up the Dance Halls." *History Today* 39, no. 10 (October 1989): 20–26.

Phelan, Peggy. *Unmarked: The Politics of Performance*. New York: Routledge, 1993.

Pitts, Victoria. *In The Flesh: The Cultural Politics of Body Modification*. New York: Palgrave Macmillan, 2003.

Polhemus, Ted, and Housk Randall. *The Customized Body*. New York: Serpent's Tail, 1996.

Pollock, Della. "Introduction: Making History Go." In *Exceptional Spaces*, ed. Della Pollock, 1–45. Chapel Hill, NC: The University of North Carolina Press, 1998.

———. "Performing Writing." In *The Ends of Performance*, ed. Peggy Phelan and Jill Lane, 73–103. New York: New York University Press, 1998.

Pomeroy, Wardell B. Foreword to *Bad Boys and Tough Tattoos*, by Samuel M. Steward. New York: The Haworth Press, 1990.

Poovey, Mary. "Creative Criticism: Adaptation, Performative Writing, and the Problem of Objectivity." *Narrative* 8 (2000): 109–133.

Rabine, Leslie W. "A Woman's Two Bodies: Fashion Magazines, Consumerism, and Feminism." In *On Fashion*, ed. Shari Benstock and Suzanne Ferriss, 59–75. New Brunswick, NJ: Rutgers University Press, 1994.

Radway, Janice. *Reading the Romance*. Chapel Hill, NC: University of North Carolina Press, 1984.

Raspa, Robert F., and John Cusack. "Psychiatric Implications of Tattoos." *American Family Physician* (May 1990): 1481–1487.

Reason, Matthew. *Documentation, Disappearance and the Representation of Live Performance*. New York: Palgrave MacMillan, 2006.

Roach, Joseph. *Cities of the Dead*. New York: Columbia University Press, 1996.

———. "Culture and Performance in the Circum-Atlantic World." In *Performativity and Performance*, ed. Andrew Parker and Eve Kosofsky Sedgwick, 45–64. New York: Routledge, 1995.

Roberts, Timothy A. and Sheryl A. Ryan. "Tattooing and High-Risk Behavior in Adolescents." *Pediatrics* 110 (December 2002): 1058–1063.

Rosencrans, Jennipher Allen. "Wearing the Universe: Symbolic Markings in Ealry Modern England." In *Written on the Body*, ed. Jane Caplan, 46–60. Princeton, NJ: Princeton University Press, 2000.

Ross, Stephen. "Introduction: Working-Class Fictions." *Modern Fiction Studies* 47 (2001): 1–11.

Rubin, Arnold. "The Tattoo Renaissance." In *Marks of Civilization: Artistic Transformations of the Human Body*, ed. Arnold Rubin, 233–264. Los Angeles, CA: Museum of Cultural History, University of California, 1988.

Russo, Mary J. *The Female Grotesque: Risk, Excess, and Modernity*. New York: Routledge, 1994.

Said, Edward. *Orientalism*. New York: Random House, 1978.

Sanders, Clinton R. *Customizing the Body: The Art and Culture of Tattooing*. Philadelphia, PA: Temple University Press, 1989.

———. "Marks of Mischief: Becoming and Being Tattooed." *Journal of Contemporary Ethnography* 16 (1988): 395–432.

Santaolalla, Isabel. "Introduction: What is 'New' in 'New' Exoticisms?" In *"New" Exoticisms*, ed. Isabel Santaolalla, 9–17. Atlanta, GA: Rodopi, 2000.

Schechner, Richard. *Between Theatre and Anthropology*. Philadelphia: University of Pennsylvania Press, 1985.

———. *Performance Studies: An Introduction*. New York: Routledge, 2002.

Schechner, Richard and Willa Appel, ed. *By Means of Performance*. Cambridge: Cambridge University Press, 1990.

Schneider, Rebecca. "Intermediality, Infelicity, and Scholarship on the Slip." *Theatre Survey* 47 (November 2006): 253–260.

Schrader, Abby M. "Branding the Other/Tattooing the Self: Bodily Inscription among Convicts in Russia and the Soviet Union." In *Written on the Body*, ed. Jane Caplan, 174–192. Princeton, NJ: Princeton University Press, 2000.

Seaton, Elizabeth. "Profaned Bodies and Purloined Looks: The Prisoner's Tattoo and the Researcher's Gaze." *Journal of Communication Inquiry* 11, no. 2 (1987): 17–25.

Sherwood, Robert Edmund. *Here We Are Again: Recollections of an Old Circus Clown*. Indianapolis, Indiana: The Bobbs-Merrill Company, 1926.

Shumway, David R. "Watching Elvis: The Male Rock Star as Object of the Gaze." In *The Other Fifties*, ed. Joel Foreman, 124–143. Chicago, IL: University of Illinois Press, 1997.

Sibley, David. *Geographies of Exclusion*. New York: Routledge, 1995.

Silverman, Kaja. "Fragments of a Fashionable Discourse." In *On Fashion*, ed. Shari Benstock and Suzanne Ferriss, 183–196. New Brunswick, NJ: Rutgers University Press, 1994.

Sims, Amy C. "Athletes Sell Skin Space to Advertisers." April 1, 2002 Foxnews.com, http://www.foxnews.com/story/0,2933,49197,00.html (accessed December 15, 2005).

Slinn, E. Warwick. "Poetry and Culture: Performativity and Critique." *New Literary History* 30 (1999): 57–74.

Sloop, John. *Disciplining Gender: Rhetorics of Sex Identity in Contemporary U.S. Culture*. Boston, MA: University of Massachusetts Press, 2004.

Sontag, Susan. *Against Interpretation*. New York: Farrar, Straus & Giroux, 1966.

Spry, Tami. "Preface to the Proceedings." In *The Green Window: Proceedings of the Giant City Conference on Performative Writing*, ed. Lynn C. Miller and Ronald J. Pelias, viii–ix. Carbondale, IL: Southern Illinois University, 2001.

Steward, Samuel M. *Bad Boys and Tough Tattoos*. New York: Haworth Press, 1990.

Stratton, John. "Building a Better Body." In *Sportcult*, ed. Randy Martin and Toby Miller, 151–172. Minneapolis, MN: University of Minnesota Press, 1999.

Sullivan, Nikki. *Tattooed Bodies*. London, Praeger, 2001.

Sussman, Henry. "Deterritorializing the Text: Flow-Theory and Deconstruction." *MLN* 115 (2000): 974–966.

Taylor, A. J. W. "Tattooing among Male and Female Offenders at Different Ages in Different Types of Institutions." *Genetic Psychology Monographs* 81 (1970): 81–119.

Taylor, Mark C. "Skinscapes." In *Pierced Hearts and True Love: A Century of Drawings for Tattoos*, ed. Don Ed Hardy, 28–45. Honolulu, HI: Hardy Marks Publications/New York: The Drawing Center, 1995.

Thomson, Rosemarie Garland. "From Wonder to Error: A Geneaology of Freak Discourse in Modernity." In *Freakery: Cultural Spectacles of the Extraordinary Body*, ed. Rosemarie Garland Thomson, 1–19. New York: New York University Press, 1996.

Turim, Maureen. "Seduction and Elegance: The New Woman of Fashion in Silent Cinema." In *On Fashion*, ed. Shari Benstock and Suzanne Ferriss, 140–158. New Brunswick, NJ: Rutgers University Press, 1994.

Turnbull, Colin. "Liminality: A Synthesis of Subjective and Objective Experience." In *By Means of Performance*, ed. Richard Schechner and Willa Appel, 50–81 Cambridge, MA: Cambridge University Press, 1990.

Turner, Bryan. "Recent Developments in the Theory of the Body." In *The Body: Social Process and Cultural Theory*, ed. Mike Featherstone, Mike Hepworth, and Bryan S. Turner, 1–35. London: Sage, 1991.

Vale, V., and Juno, Andrea. *Modern Primitives: An Investigation of Contemporary Adornment and Ritual*. San Francisco, CA: Re/Search Publications, 1989.

Wallis, Brian. Introduction to *Art After Modernism: Rethinking Representation*, ed. Brian Wallis, xi–xviii. New York: New Museum of Contemporary Art. Boston: Godine, 1984.

Wegenstein, Bernadette. "Getting under the Skin, or, How Faces have Become Obsolete." *Configurations* 10 (2002): 221–259.

West, Candace, and Don H. Zimmerman. "Doing Gender." In *The Social Construction of Gender*, ed. Judith Lorber and Susan A. Farrell, 13–37. London: Sage Publications, 1991.

Wiley, Stephen Crofts. "Spatial Materialism: Grossberg's Deleuzian Cultural Studies." *Cultural Studies* 19 (2005): 65–99.

Williams, Raymond. "Advertising: The Magic System." In *The Cultural Studies Reader*, 2nd edition, ed. Simon During, 410–426. New York: Routledge, 1999.

Wise, J. MacGregor. "Assemblage." In *Gilles Deleuze: Key Concepts*, ed. Charles J. Stivale, 77–87. Ithaca, NY: McGill-Queen's University Press, 2005.

INDEX

CPSIA information can be obtained at www.ICGtesting.com
Printed in the USA
LVOW12*2125160713

343245LV00004B/44/P

9 780230 600270